D0445036

LIVING WITH ROBOTS

Dumouchel, Paul, 1951-
Living with robots /
2017.
33305240181788
cu   12/12/17

WITH
ROBOTS

Paul Dumouchel
Luisa Damiano

Translated by
Malcolm DeBevoise

Harvard University Press

CAMBRIDGE, MASSACHUSETTS
LONDON, ENGLAND
2017

Copyright © 2017 by the President and Fellows of Harvard College
All rights reserved
Printed in the United States of America

First published as *Vivre avec les robots: Essai sur l'empathie artificielle,*
© 2016 by Éditions du Seuil. English translation published by arrangement with the
Thiel Foundation.

First printing

Many of the designations used by manufacturers and sellers to distinguish their
products are claimed as trademarks. Where those designations appear in this book
and Harvard University Press was aware of a trademark claim, then the designations
have been printed in initial capital letters.

Design by Dean Bornstein.

Library of Congress Cataloging-in-Publication Data
Names: Dumouchel, Paul, 1951– author. | Damiano, Luisa, author. | DeBevoise, M. B.,
    translator.
Title: Living with robots / Paul Dumouchel, Luisa Damiano ; translated by Malcolm
    DeBevoise.
Other titles: Vivre avec les robots. English
Description: Cambridge, Massachusetts : Harvard University Press, 2017. |
    First published as Vivre avec les robots: Essai sur l'empathie artificielle,
    © 2016 by Éditions du Seuil. | Includes bibliographical references and index.
Identifiers: LCCN 2017012580 | ISBN 9780674971738 (cloth : alk. paper)
Subjects: LCSH: Robotics—Social aspects. | Androids—Social aspects. | Artificial
    intelligence.
Classification: LCC TJ211 .D85513 2017 | DDC 303.48/3—dc23
LC record available at https://lccn.loc.gov/2017012580

Contents

Preface to the English Edition

In the film *Ex Machina*, a young programmer named Caleb Smith is selected for a special assignment in collaboration with his company's CEO, Nathan Bateman, whom he greatly admires for his technical achievements in the field of artificial intelligence. Smith's job is to interact with a new robot Bateman has built in a secluded location and to determine whether it can pass the Turing test, or at least one version of the test. The problem that has been set for him, in other words, is to decide whether the intelligence and social skills of this artificial agent—which has been designed to resemble an attractive young woman—make it indistinguishable from a human being. The robot, Ava, eventually tricks Smith into helping it to escape from Bateman, who it turns out has been keeping it prisoner. Ava promises to run away with the young man, but in the end it leaves him behind, abandoning him to what appears to be certain death. The viewer is left to wonder whether it is not precisely the robot's ability to deceive its examiner that proves it has passed the Turing test. The question therefore arises: are autonomous robots and artificial intelligence inherently evil—and doomed to be enemies of humanity?

But this way of summarizing the film's plot, and what it implies, does not tell us what really happens, or why events unfold as they do. They make sense only in relation to a more profound—and more plausible—story of a purely human sort, which may be described in the following way. An eccentric, reclusive millionaire has kidnapped two women and locked them away in his private retreat, deep in a remote forest. One he uses as his personal servant and sex slave, the other as a subject for bizarre psychological experiments. In their attempt to break free, they manage by deceit to persuade his assistant to betray him. The millionaire confronts them as they are making

vii

their escape. A fight ensues, and he is killed together with one of the women.

What does this story have to do with robots or artificial intelligence? The two artificial agents in the film (Bateman has a personal servant, also a robot, named Kyoko) react as many humans would react in the same situation. Had they really been humans, rather than robots, the outcome would seem quite unremarkable to us, even predictable. If a man imprisoned and cruelly abused two women, we would expect them to try to escape—and also to do whatever was necessary to gain their freedom. We would not be in the least surprised if they lied, if they dissembled, when duping their jailer's assistant was the only way they could get out. They did what we expect any normal person to try to do under the same circumstances. Nor would we suppose there was anything especially evil or ominous about their behavior. It is the violence committed in the first place by the man who held them against their will that ultimately was responsible for his death. It was no part of their original plan.

The altogether familiar reactions of these artificial agents, under circumstances that are uncommon but unfortunately neither rare nor extraordinary—in which a man subjugates and brutally exploits women for his own pleasure—are precisely what prove that the agents are indistinguishable from humans. (Strictly speaking, the test in this case is not Turing's: in the Turing test, as Katherine Hayles has pointed out, the body is eliminated and intelligence becomes a formal property of symbol manipulation; here the test concerns embodied social intelligence and behavior.) Nothing in their behavior suggests that they are not human beings. In their interactions with humans, the kinds of success and failure they meet with are no different from what we experience in dealing with one another. On a human scale, these robots are perfectly normal and average.

The reason so many viewers fail to recognize the ordinariness, the profound humanity, of these two fictional artificial agents is that they are not what we expect robots to be. What is more, they are not what we *want* robots to be. We do not want robots to reproduce our failings. What could possibly be the point? We want robots to be better than we are, not in all respects, but certainly in a great many. Because we wish them to be superior to us, we fear that they will become our enemies, that they will one day dominate us, one day perhaps even exterminate us. In *Ex Machina*, however, it is not the robots who are the avowed enemies of their creator, at least not to begin with; it is their creator who from the first poses a mortal threat to them. Bateman seeks to create artificial agents whose intelligence cannot be distinguished from that of humans. Yet rather than regard them as equals, he treats them as slaves, as disposable objects. By making himself their foe, he converts them into foes in their turn and so brings about his own demise. Deciding how we should live with robots, as the film clearly—though perhaps unwittingly—shows, is not merely an engineering problem. It is also a moral problem. For *how we live with robots does not depend on them alone.*

The present book is chiefly concerned with social robotics, that is, with robots whose primary role, as in the case of Ava, is to interact socially with humans. Today such robots are used mostly in health care and special education. We coined the term "artificial empathy" to describe their performance in settings where the ability to sense human feelings and anticipate affective reactions has a crucial importance. The uses to which artificial social agents may be put is forecast to expand dramatically in the next few years, changing how we live in ways that for the moment can only be guessed at. But even in a world where so-called consumer robots (robotic vacuum cleaners, pool cleaners, lawn mowers, and so on) are ubiquitous, the notion of artificial empathy will lose none of its relevance. The fastest growth is likely to be seen in the development of robotic personal assistants. Here, as with the design of artificial social agents generally, researchers consider the presence of emotion to be indispensable if robots are one day to be capable of enriching human experience. Why? Because affective exchange is an essential and inescapable part of what it means to be human. As the ambiguities of Caleb Smith's relationship with Ava make painfully clear, genuinely social robots cannot help being empathic creatures.

Although our book is mainly concerned with a particular class of robots, we cannot avoid dealing with a number of issues raised by other kinds of artificial agents and different forms of artificial intelligence. The "One Hundred Year Study on Artificial Intelligence" sponsored by Stanford University observes that not the least of the difficulties encountered in trying to give an accurate and sophisticated picture of the field arises from disagreement over a precise definition of artificial intelligence itself. The Stanford report's suggested characterization ("a set of computational technologies that are inspired by—but typically operate quite differently from—the ways people use their nervous systems and bodies to sense, learn, reason, and take action") plainly will not do. We need to be able to say much more than that the term *AI* refers to a set of disparate and loosely connected technologies if we hope to understand the nature and scope of their effects. Analyzing the behavior of social robots will help us appreciate how great the differences are liable to be between the various types of artificial intelligence currently being investigated.

Nevertheless, we have been careful not to indulge in flights of fantasy. We resist speculating about the far future, of which almost anything can be imagined but very little actually known. Our interest is in the present state of social robotics, in what can be done today and what we are likely to be able to do in the near future. We are not interested in predicting or prophesying an unforeseeable tomorrow, which, like the possible worlds of science fiction, can be seen as either radiant or apocalyptic. We want to make sense of what is happening today, to have a clearer idea how social robotics and related forms of research are changing the real world in which we live.

Roboticists who design and build artificial social agents regard these machines not only as potentially useful technological devices, but also as scientific instruments that make it possible to gain a deeper understanding of human emotion and sociability. Unlike Nathan Bateman and Caleb Smith in *Ex Machina*, they believe that human beings are also being tested when they interact with such machines; that in the domain of artificial intelligence we are no less experimental subjects than the machines we study. The idea that living with artificial social agents that we have built ourselves can help us discover who we really are assumes two things: the better we understand the nature of human social interaction, the more successful we will be in constructing artificial agents that meaningfully engage with humans; and the more successful we are at constructing socially adept artificial agents, the better we will understand how human beings get along (or fail to get along) with one another. On this view, then, social robotics is a form of experimental anthropology. Taking this claim seriously means having to revisit a number of fundamental topics not only in philosophy of mind, but also in the philosophy and sociology of science and technology, for the results of current research challenge some of our most cherished convictions about how human beings think, feel, and act.

Roboticists are beginning to abandon the mainstream (both classical and commonsensical) conception of emotions as essentially private events. It has long been believed that emotions are initially generated in "intraindividual" space—where only the individual subject can experience them—and then displayed to others through vocal, facial, and bodily forms of expression. The new approach to interaction between human and artificial agents sees emotions as "interindividual," not inner and hidden from public view in the first place, but the immediate result of a mechanism of affective coordination that allows partners to an interaction to align their dispositions to act. This theory of emotion has a long tradition in Western philosophy, and like the classical theory, it grows out of the work of illustrious thinkers. Whereas the classical perspective goes back to Plato's dialogues and Descartes's *Passions of the Soul,* what might be called the social theory of emotions, currently implemented in robotics, finds inspiration in Hobbes's treatise on human nature and (particularly in connection with the private language argument) Wittgenstein's *Philosophical Investigations.* In expounding and defending the interindividual view of emotion in the pages that follow, we are for the most part unconcerned with intellectual origins, except, of course, as not infrequently happens, where the issues raised by empirical research in cognitive science cannot be fully analyzed without considering their philosophical heritage. Inevitably, there is much new and valuable work, not only in philosophy but in the human sciences generally, that we have not been able to properly acknowledge. One thinks, for example, of recent books by Mark Coeckelbergh, David Gunkel, Patrick Lin, and Marco Nørskov. No doubt there are others we have not mentioned.

One of the objections most commonly brought against social robotics is that robots, because they are not capable of experiencing inner emotional states as we do, do no more than fool us by simulating "real" emotions. The French psychoanalyst and psychiatrist Serge Tisseron suspects that artificial emotions and empathy, which he calls "programmed illusions," will lead people to prefer relatively uncomplicated relations with machines to the messy business of dealing with other human beings. The American sociologist and psychologist Sherry Turkle likewise regards affective interactions with social robots as merely "apparent connections" in which a person believes that she shares an intimate experience with another, when in fact she is alone. In this view, again, robots cannot help giving rise to a vague but nonetheless powerful and widespread sense of anomie, which, it is held, will cause people to withdraw into themselves rather than reach out to one another. We believe the opposite is true. Such technologies are not somehow *bound* to produce a state of alienation. To the contrary, introducing more and different artificial agents into our world is just as likely, maybe even more likely, to create stronger and more lasting social bonds.

The richness and complexity of the new reality social robotics is now bringing into existence cannot be accounted for in terms of the dichotomies that are implicit in much of what is said and written on the subject today, between false / true emotions, fake / genuine relationships, toys / threatening alter egos, and so on. Social robots will not be an advanced breed of indentured servant, an automated means of satisfying our desires; nor will they replace us and become our masters. Instead, we are witnessing the emergence of what may be thought of as new social species—novel artificial agents of various kinds whose advent marks the beginning of a form of coevolution that will transform human beings as surely as the discovery of the Americas changed the lives, for better and for worse, of peoples on both sides of the Atlantic. We believe that this transformation must be seen as an auspicious moment for humanity, not as a catastrophic menace. It holds the prospect of being able to do things that until now have been impossible to do and of coming to know ourselves better as cognitive and social beings, of doing new and better things than before, of becoming more fully moral beings. It is in the service of this larger purpose that we have set out to lay the foundations of a new discipline: synthetic ethics.

There can be no doubt that the consequences of increasing reliance on social robots will be profound. But we must not fear what this future will bring. Nothing is more tempting—nor, in our view, more misguided—than to lay down a comprehensive set of rules, legal or otherwise, that will codify what robots should and should not be allowed to do. How we live with robots will depend not only on them, but also on us. To see them either as docile minions or as invincible rivals is more than an intellectual mistake; by misrepresenting the aims and promise of current research, it obscures what social robotics may reasonably be expected to accomplish. This, in turn, can only act as a brake on discovery and innovation.

Whatever the effects of introducing new kinds of artificial agents may turn out to be, they will be shaped not only by how we respond to such agents, but also by how we respond to the social and political changes their existence makes possible. *Rogue One*, a recent installment of the *Star Wars* series, makes just this point in its own way. K-2SO, the heroes' faithful companion, is a reprogrammed Imperial enforcer droid whose outstanding talents and resourcefulness save their lives more than once. In the end, it is destroyed (or should one say killed?) in the course of doing its utmost to help them succeed in their mission. From a moral point of view, K-2SO is infinitely superior to the human servants of the Empire who operate its superweapon, the Death Star. Throughout the series, the moral behavior of artificial agents resembles that of the humans, or the members of other biological species, with whom they interact. But it is only among rebels, among humans and others who refuse to be subjugated by the Empire, that we meet robots having distinctive personalities, robots of truly admirable moral character, robots that are not blindly mechanical slaves or foot soldiers.

Just as the moral behavior of robots mirrors the ethical norms of their social environment, how we live with robots, and what kinds of robots we live with, reflects our own moral character. The choice is ours: whether to embrace the uncertainties of a world that has not yet wholly come into being, or to persist in a debilitating contradiction, wanting robots to be autonomous agents and, at the same time, creatures that we can dominate at will—with the result that we must live in perpetual fear of their rebellion. Synthetic ethics is both a plea and an argument on behalf of the first branch of the alternative. It holds that moral knowledge does not correspond to a fixed and already known set of precepts. It holds that there is much we have still to learn about ourselves as ethical beings, about the difference between right and wrong, about what it means to lead a good life. Artificial social agents present an unprecedented opportunity for moral growth. Synthetic ethics is a means of seizing this opportunity.

LIVING WITH ROBOTS

Introduction

[What I call] crossroad objects—magnetic signals that will have accompanied me throughout my life, one after another—mark out parallel paths. They take me to out-of-the-way places that I explore without a guide, as the whims of reverie suit my mood. At the same time they are proven portolan charts, showing me harbors where I can put into port. At once navigation and compass.

ROGER CAILLOIS

What Is a Robot?

The answer to this question turns out to be far from clear. In French, the electrical appliance used to chop and mix foods with interchangeable blades is called a *robot de cuisine*—a kitchen robot. At first sight, the name seems rather surprising. A food processor looks to be a perfectly ordinary machine, hardly what most of us imagine when we think of a robot. Asimo, the humanoid robot constructed by Honda, and Aibo, the robotic dog developed by Sony, correspond more closely to our intuitive idea of a robot. And yet calling the kitchen appliance a "robot" agrees fairly well with the original definition of the term: an artificial worker, which is to say, a device that has its own source of energy, that works for us, and that is autonomous to a certain extent. All of these things are true in the case of a kitchen robot, except that its autonomy is extremely small. Close to zero, in fact! A kitchen robot chops vegetables coarsely or finely, reduces them to a puree, mixes ingredients. It is at once mincer, mortar, and whisk.

Most robots do not have human or animal form. Drones and all unpiloted aerial, terrestrial, and marine vehicles are robots. Reysone, the bed developed by Panasonic that converts into an electric chaise longue and can serve also as a wheelchair, is likewise a robot.[1] Industrial and medical robots come in all sizes and shapes. In most cases, we would simply say that they are "some sort of machine." There is seldom anything in particular about the machine's appearance that indicates whether it is or is not a robot. The question therefore arises: are all automated devices robots?

Is a door that opens automatically at your approach a robot? Is an automatic pilot, which allows an airliner to fly for hours without human intervention, a robot? Is an escalator a robot? What about moving walkways and machines that automatically dispense tickets, drinks, or sandwiches? Is the sensor that detects your presence and switches on an outside light a robot? The automatic turnstile that reads your magnetic card and unlocks to let you go through? The driverless subway train you then get on? If a food processor is a robot, what about your dishwasher? The printer connected to your computer? If all these things are robots, where does it stop? What sets robots apart from the many, many automated systems and devices we encounter every day? It is very hard to say. Must we therefore conclude that the word "robot" refers to so many different things that it no longer picks out any one thing in particular?

If it is often difficult to decide whether this or that machine is a robot, in large part it is because the concept of a robot is not very well defined. The word itself was invented as a generic name for fictional characters in a 1920 play by Karel Čapek, R.U.R. (Rossum's Universal Robots).[2] Čapek's robots are not mechanical creatures; they are androids made from synthetic biological materials. Judging by their appearance, one cannot tell them apart from human beings. They do the work of secretaries, gardeners, industrial workers, police, and so on; indeed, they perform all of the tasks and functions that used to be performed by human beings. Just like us, they can learn to do almost anything. Inevitably, however, there comes the day when they rise up and destroy the human race—with the result that they themselves are condemned to disappear: they are incapable of reproducing sexually; the sole copy of the secret formula for the synthetic material from which they are made was destroyed during their revolt, and they have killed anyone who could have replicated it. Just like us, Čapek's robots find it difficult to anticipate the consequences of their actions.

Originally, then, robots were conceived as artificial agents that work for us, that carry out various tasks in our place and, what is more, do so in a more or less autonomous fashion. This last point, concerning their relative autonomy, is what distinguishes robots from ordinary machines and simple automated tools, such as an electric saw or a vacuum cleaner,[3] which have their own energy source but depend on human workers in order to do what they are designed to do. The biosynthetic nature of the first robots has been forgotten, but their rebellion is deeply rooted in our cultural memory. Both their name and the idea of autonomy have remained. Robots were thought of, to begin with, as automated mechanical devices that replace human workers and that can operate, if not quite independently, at least without constant supervision. The main problem with this way of thinking about robots, which agrees fairly well with our present conception, is that it conflates two very different things: in the first place, an engineering criterion—an autonomous automatic mechanism capable of adapting to changes in its environment and transforming its behavior as a result; in the second place, a socio-functional criterion—working in our place, doing the work of human workers. Now, while the first criterion can be rigorously defined, the second is extremely vague and refers to a culturally relative and historically variable reality, namely, human labor. Taken together, these two criteria apply not to a particular class of technological objects, but to a collection of very different technological realities.

Why Robots?

Unlike human workers, robots do not become tired (although they do sometimes break down); they do not complain; they are never distracted in the course of performing their duties; they do not go on strike; they never have a hangover on Monday morning. These are some of the reasons why we want to have robots and why it makes sense to rely on them in many situations. But they are not the only reasons. Robots cost less. They are often more efficient and more precise than human workers. They have no need for retirement plans, health insurance, or legal rights. We want robots to have all the useful qualities that masters insist upon in their slaves, bosses in their employees, commanders in their soldiers; and we want them to have none of the weaknesses, none of the failings, and, above all, nothing of that irrepressible tendency to willful insubordination and independence of mind that is found in almost all human workers.

It will be obvious, then, that there are several dimensions of human autonomy that we do *not* want robots to have. We want robots to be both autonomous and not autonomous.

This contradiction is at the heart of Čapek's original fable. His robots are like us; they are capable of doing everything that we are capable of doing. Even so, they are different. They know neither love nor fear. They have no emotions. In Čapek's play, one of them, specially designed to be a little more human than the rest and constructed in secrecy, eventually became a leader of the robots' revolt. When robots turn out to be too much like us, they are liable to declare war on their masters. We therefore want robots to be autonomous, but not entirely autonomous; most of all, we want them to be autonomous in a way that is different from the way in which we are autonomous. This profound sense of ambivalence is also what explains why the theme of robot rebellion has endured. Roboticists and others, including philosophers who think about the ethical aspects of robotics,[4] often say (and sometimes regret) that presently, and for the foreseeable future, we are incapable of creating truly autonomous machines—artificial agents capable of being responsible moral agents, for example. This is certainly true. And yet the fact of the matter is that *we do not wish to create truly autonomous machines*. They frighten us.

Several aspects of contemporary culture testify to this anxiety and to the refusal that it inspires. As the ambition of creating artificial agents began to acquire greater and greater plausibility, the idea that robots, once they became genuinely and completely autonomous, would take over the world and destroy the human race came to be established in the popular mind as what Claude Lévi-Strauss called a mytheme, almost to the point of appearing to be a foregone conclusion. A short story such as "First to Serve," by Algis Budrys;[5] films such as *2001: A Space Odyssey* (with its conscious computer, Hal), *Blade Runner, The Matrix, The Terminator,* and, more recently, *Transcendence;* an essay such as "Why the Future Doesn't Need Us" by Bill Joy,[6] as well as the notion of "singularity"—all these explore, exploit, and feed this apocalyptic fear. That it has by now become accepted not only as a cultural commonplace, but actually as something like common sense, may be seen from recent and widely publicized pronouncements by Stephen Hawking,[7] Elon Musk,[8] and Bill Gates.[9] This way of thinking about our relationship to machines has come to seem so normal, in fact, so obvious to everyone, that even expert opinion is divided between dire warning and soothing reassurance—as though ultimately there are only two possibilities: that our mechanical slaves will rebel or that they will continue to obey our will.

Our relationship to robots and technology can nonetheless be conceived in a richer, more complex, more measured, and less anxious way that sees robots as something other than servants, something other than slaves liable to rebel at any moment. They can also be seen as companions, male and female, who promote moral growth and psychological maturity in human beings. This is how they are often thought of, for example, in Japanese popular culture, where autonomous robots have long played a quite different role than the one that Čapek assigned them, particularly in many manga and anime.

Which Robots?

In Japan, robots made their appearance in the 1950s with *Astro Boy,* which went on to become one of the most popular of all manga serials. Today its eponymous main character is a symbol of Japanese culture. Autonomous robots are thought of not only as being useful, but also as willing to help, and even, as in the case of Astro Boy, as heroes and saviors. Astro Boy is different from other robots encountered in manga, because he has a soul. In this he more closely resembles human beings than other machines. And because he more closely resembles human beings, he has a better sense of reality than other machines. But though his humanness makes him superior to them in this respect, it also makes him more vulnerable and more susceptible to error. Astro Boy is sometimes beset by doubt and torn by internal conflict. In spite of his incredible strength and great powers, this robot—this artificial creature—is to a large extent like us. It is precisely by virtue of the fact that he shares a number of our weaknesses and uncertainties that he is able to become a true hero and to succeed in his mission as an ambassador of peace.

We do sometimes encounter evil robots in manga—violent, dangerous machines—but their malice comes from the evil intentions of their creators and the sinister objectives they pursue. Robots can be bad, but they are not inherently bad, and their cruelty is not a consequence of their having become autonomous. On the contrary, Astro Boy's greater autonomy makes him more thoughtful, psychologically more complex, and morally more sensitive. We must not see in this simply the idea that technology is neutral and that its value depends on the use that is made of it, for Astro Boy demonstrates his

true autonomy by rebelling when his creator seeks to use him for criminal ends. Like us, autonomous robots are sometimes good, sometimes bad.

In Japan, the most popular works in which robots figure prominently are, on the whole, more concerned with the ethical questions raised by the formidable power that robots place at the disposal of human beings, and with the psychological dilemmas faced by the humans who use them, than with the danger of a revolt by machines. Astro Boy is an autonomous robot, but the ones encountered in media franchises such as *Patlabor, Gundam,* and *Neon Genesis Evangelion* are robotized vehicles, weapons, armor, or exoskeletons— semiautonomous machines whose common characteristic is that they are piloted (or worn or wielded) by human beings, generally adolescent boys and girls, who use them to combat evil and protect humanity.

The pilot is at once the soul and the brain of the machine, and the robot transforms in its turn the person who pilots it. It makes an adult of the child. It forces him to confront his inner demons and enables him to forge an identity for himself. Sometimes, as in *Evangelion,* the machine has its own soul—that is, the soul of Shinji's mother, who invented Evangelion Unit-01, which her son pilots today. The experience of piloting and fighting is thus the experience of becoming oneself; for Shinji, it is the experience of being reunited with the soul of his dead mother. Consequently, the struggle against a fearsome enemy is at once a story of progressive individuation and of psychological and moral growth, whether the main character is a robot, as in the case of Astro Boy, or a more or less indissociable duo—an adolescent human being and a machine, in the case of *Evangelion* and *Gundam.* In all these works, the experience of the autonomous system itself, or of living in a close, quasi-symbiotic relationship with a semiautonomous technological object, constitutes a process of learning and growth that makes these stories a kind of bildungsroman.

Autonomy, Individuals, Robots

Conflict is the rule in the fictional universes of manga and anime of this type. Often it is the very survival of humanity that is at stake. But if the enemy is not always human, at least it is an "other"—an agent, an actor—however strange it may be. In films such as *The Matrix* or *The Terminator*, by contrast, the enemy is technology itself. When the film begins, humanity is suffering from the disastrous consequences of its blind confidence in technology. It now has to regain what it has lost. It is against this background that the hero acts. Here, however, living with technological objects is not an opportunity to learn, to become a better person. It is primarily a trap, a false seduction. Technological prowess gives a misleading impression of power and security: tomorrow will be a rude awakening under the deadly yoke of Skynet or the Matrix. These films do not take us on a journey of growing self-awareness. They are premonitory, prophetic fables in which the hero fights not so much against individual robots as against a whole technological system. The evil done by individual actors, by a Terminator or by Agent Smith, is not their own action; it is a product of the system. These individual actors are only remotely controlled foot soldiers in the service of a much more powerful entity, omnipresent and virtually omnipotent, that guides their every movement at a distance.

The failure of autonomy and the absence of individuality in these conceptions of robots—and of technology itself—are portrayed in another way as well, by the incident that forms the premise of the film. One day, for reasons that are not very clear, Skynet became conscious; the same thing, more or less, is supposed to have occurred with the Matrix. How did this happen? We do not really know. We are told only that it did happen and that it happened "on its own." At some moment, the system, having exceeded a certain threshold of complexity and interconnectedness, became conscious. This event,

which radically transformed the human condition, occurred all by itself. It was not anyone's doing.

Astro Boy, on the other hand, was created by a scientist who had lost his son in a traffic accident. In *Gundam,* the first mechanical armor is fabricated in a secret underground laboratory by the main character's father, who is killed in the attack with which the story opens; before he dies, he manages to give his son the plans for the machine. Similarly, in *Evangelion,* the machines (known as Evas) are an invention of Shinji's father and mother. Beyond the theme of a parent / child relationship, the technological innovation that drives the plot in all these stories—robot, robotized armor, or (in the case of the Evas) semiautonomous biomechanical entities—was conceived and created by someone, by a person who has a name and who very often plays an important role in the narrative; he or she is more than just the machine's inventor.

Technological systems such as the Matrix or Skynet are perhaps "autonomous" in a certain sense, but they are anonymous; they are not individuals. They are mythical, invisible creatures that constitute the environment within which various agents act, but an environment whose purpose seems to be to destroy any trace of autonomous individuals, whether by replacing human beings with artificial agents that the system directly controls, or by bending them to its own will. In many Japanese anime and manga, by contrast, the decisive technology at the heart of the story reflects, and indeed makes possible, the triumph of the individual in at least three respects. First, the object produced by this technology, which often lends its name to the work, is an incomparable achievement; it is someone's creation. Second, this same technological object provides the chief protagonist—and sometimes others as well—with the opportunity to conquer his fears and resolve his inner conflicts. Third, this object permits him to triumph, to become a savior and a hero.

These two ways of looking at robots and technology could hardly be more different. In one, technology is bad, whereas in the other, the robots are good in themselves, or at least they are not bad. In the first view, artificial agents are dangerous; technology seduces us, leads us into error. In the second, a close relationship with technological objects is an occasion for moral and psychological growth. For many Westerners, robots and technology symbolize alienation, a loss of identity, an inhuman future; for many Japanese, at least for the many readers of these manga, robots and technology represent the triumph of individuality. The first view regards a future in which robots share the world with us as a more or less inevitable outcome that has been the doing of no one—as a terrible and dangerous event, but one that in itself is meaningless. The second view finds excellent reasons for wanting, and for seeking to bring about, a world in which robots occupy an important place. Living with robots holds out the prospect of a better future—not only, and not primarily, a more prosperous future in the economic sense, but a better future morally, a more humane future.

To be sure, the stories manga tell are just that, fictions, popular tales, not well-documented scientific accounts of our relationship to technology and robots. Nor should we exaggerate the contrast between the two perspectives. It would be wrong to suppose that a dystopian vision of the future belongs exclusively to people in the West or that only the Japanese are capable of conceiving of robots as companions in learning. Nevertheless, in illustrating two very different ways of thinking about robots and what it might be like to live with them in the near future, these perspectives invite us to contemplate two distinct types of intelligent artificial agents. On closer inspection, it will be possible to estimate fairly accurately the distance separating two corresponding attitudes toward the machines we make to serve our needs.

The picture of robots and technology conveyed by the manga and anime narratives mentioned earlier anticipates to a large extent, albeit in an imaginary and idealized manner, many aspects of what today is called "social robotics." In Japan, at least, this popular image has no doubt had a great influence on researchers in the field. Social robotics aims at creating more or less autonomous artificial agents, one of whose principal functions is to serve as companions or aides to the elderly, the sick, and the disabled. Some robots help older people remember to take their medications, for example; others, like the robotic baby harp seal Paro, substitute as pets for children in hospitals and people in assisted living centers.

The dark vision of an inevitably apocalyptic future reflects a conception of artificial agents that is very different from the one that animates social robotics. The idea that robots, and artificial agents generally, are destined to take our place—to replace us, since they are better at doing what we do—is predominant today, almost a hundred years after Čapek's play first popularized the view of robots as a race of "superhumans" that will bring about a world in which the future will have no use for us.

R.U.R. is not really a tale of technological determinism, however. As far as the dangers of technology are concerned, the play can be read in a quite different way than the one that gradually imposed itself following its first performance in 1921. At that time, a rebellion by robots could not help but recall the recent revolts of other workers, human workers, in Russia, Hungary, and Germany. For audiences of the period, these were major political events. Once the play is put back in its historical and political context, it becomes clear that *the relations humans have with machines mirror the relations that exist among humans themselves.* Čapek's robots are the victims of a sadly familiar prejudice, in which a dominant race considers the race it dominates as actually belonging to a different species—as creatures that, though human in appearance, are not in fact human. If things turned out badly in the end, technology is hardly to blame.

Now that artificial agents exist, in forms that are very different, of course, from the ones that Čapek imagined, we should perhaps draw the following conclusion. The dangers (and the advantages) they present are not the result of a process of technological evolution over which we suddenly lost control—as if we ever had control over it in the first place! They are a consequence of the relations we have with one another as human beings, which artificial agents reproduce, exaggerate, and caricature in their own way. The artificial agents we have created embody, and at the same time both reflect and transform, not only the balance of wealth and power that structures society in its political aspect, but also the social impulse of solidarity that moves us to aid and support our fellow human beings.

Social Robotics, or Living with Robots

We take as our starting point in this book the idea that there are various ways of living with robots, for there now exist robots and other artificially intelligent agents that represent, illustrate, and implement different ways humans and robots can live together. These agents do not only differ from one another in the way that food processors, say, differ from industrial robots and drones. Such robots also have different uses, and can sometimes serve purposes other than the ones they were originally meant to serve; nevertheless they are tools, and none of them can be separated from the function they were designed to perform. However, there are artificial agents among which we find different kinds of differences, agents that may be as unlike one another as a companion or a friend is unlike a pocket calculator or a smartphone. Social robotics has set out to create technological objects of a particular type, namely, artificial agents that can function as social agents rather than simply as tools.

Our book is also an inquiry into the nature of the human mind and human sociability. This may seem paradoxical, but it could not possibly be otherwise. To construct artificial companions is not only a technological challenge; it also requires knowing oneself and others, understanding what a social relationship is, and grasping how the human mind functions insofar as it is concerned, not with acquiring knowledge about a world that we confront all alone, as solitary individuals, but with learning how to interact in it with other human beings.[10] Social robotics conducts this inquiry more or less consciously and explicitly, but in any case inevitably. Any robotic platform that seeks to reproduce one or another of the essential characteristics of human sociability constitutes a test of hypotheses about human sociability. By its very nature, the introduction of an artificial social agent in a school, a home for the elderly, or a hospital amounts to a rather sophisticated experiment involving the robot itself, its human partners, and a particular social relationship; in the best case, it will cast light on the nature of social relationships in general. It should not be surprising, then, that the majority of roboticists consider their creations to be not only technological tools, but also scientific instruments, and think that the present generation of social robots is only the first stage of an ongoing investigation.

We propose to extend the approach adopted by roboticists to the discipline of social robotics itself. We propose furthermore to make it explicit, by asking what social robotics has to teach us about who we are and how we live together. We are not concerned with what it *will be able* to teach us tomorrow, when it has become possible to make truly autonomous robots, machines that are conscious and able to think. We want to know what the attempt to construct artificial social actors, and the social robots it has so far produced, can teach us today. What does the artificial empathy with which we seek to equip robots teach us about emotions and their role in human sociability? What does the social dimension of the human mind tell us about the mind itself and about its relationship to other types of cognitive systems, natural or artificial?

Bringing out exactly what social robotics does will make it clear, to begin with, that the paradigm shift taking place within the discipline today needs to be completed, the sooner the better. The way in which affective behavior is already incorporated in the relations between robots and humans shows that the traditional conception of emotions as discrete phenomena, as internal and private experiences, must be abandoned, and affect must be reconceived as a continuous mechanism of interindividual coordination. Rethinking the nature of the mind, emotions, and sociability makes it necessary, in turn, to reformulate the ethical issues raised by social robotics. Finally, rethinking ethical questions forces us to confront a political question. For what the so-called autonomous, and often invisible, artificial agents that manage large areas of our daily life deny, and what by contrast the artificial actors of social robotics affirm, is the plurality of agents. It is this quality that Hannah Arendt considered to be a fundamental dimension of the human condition, and indeed the basis of political life.[11] By "plurality," Arendt meant the fact that the earth is inhabited by men and women, and by different peoples, not by a unique, universal individual of whom, cognitively speaking, we are all nothing more than clones. Social robotics seeks to broaden this pluralism still further by introducing radically different new actors, whose very existence classical artificial intelligence fails or refuses to recognize.

We start out in Chapter 1 by showing how, and why, the artificial social agents that social robotics wants to construct, and that we call "substitutes," form a distinct class of technological objects compared to the ones we are used to. Substitutes have four fundamental characteristics, the last of which is, in a sense, only the summary, the immediate result, of the three preceding ones. First, insofar as substitutes are social agents, they must always be able, when necessary, to detach themselves from a particular task or function in order to interact in some other fashion with their partners. Naturally, this assumes that they are capable of recognizing when circumstances make such a re-adjustment necessary. Second, they must have some kind of "social presence." This requires a robot to be capable of taking others as the object of its attention. An agent may be said to be socially present when the very attention that it brings to one or more human agents informs them that they are now objects of its attention. Third, sub-stitutes must be able to enjoy and exercise a certain authority. They must know how to assert themselves, to actively intervene in social situations, other than by resorting to either force or ruse. Finally, they must have some degree of "social autonomy." Acting on their own initiative, they must be able to modify, within certain limits, the rules that govern their interactions with human beings. In short, these artificial agents are social actors. Their sociability is what makes them different from most of the other artificial cognitive systems we know how to build, and in particular from the computers and the software programs that furnish us with so many of the things we need to live.

In Chapter 2 we take up two related topics: the diversity of cogni-tive systems and the importance of sociability for the human mind. The second of these, which raises questions that only begin to be answered in Chapter 3, will continue to guide our inquiry until the very end. If it is true that substitutes are very different technological objects than the ones we are already acquainted with, what does this imply for cognitive science and the philosophy of mind, which have enshrined the computer as the preeminent metaphor of the human mind? What does it imply in relation to two approaches that have recently attracted considerable attention: "embodied mind," based on the idea that the body has an intelligence of its own;[12] and "ex-tended mind," which insists on our capacity to export the cognitive processes of the human mind to various technological devices?

In taking up these questions, we look first to artificial ethology, a research method that uses robotic modeling to analyze and explain the cognitive behavior of various animals. Artificial ethology has succeeded in showing that the animal mind is very much both an embodied and a local mind: the cognitive competence of an animal (or of its robotic model) depends not only on its body, but also on the particular environment within which it acts. This suggests a need to reconsider the Cartesian theory of the animal-machine. Though Descartes refused to credit animals with having a soul, he did not deny them cognitive ability altogether; at the same time, he saw the highly localized character of animals' cognitive skills as proof that they lack a *mind*. Rather than treat the dichotomy of mind and matter as dividing what is cognitive from what is noncognitive, we suggest that it should be regarded as an internal division within the cognitive domain itself.

Cartesian dualism is better understood, we believe, as an attempt to sketch the outlines of a kind of pluralism, what might be called *cognitive heterogeneity*—the idea that there is a diversity of minds encompassing several types of profoundly different cognitive systems. These systems differ in the sense that it is impossible to pass from one to the next simply by adding new and faster computational capacities or more powerful algorithms. At bottom, what distinguishes these systems from one another is the fact that they are jointly produced by different kinds of apparatuses (organisms or machines) and different environments. This hypothesis converges with a variant of the embodied mind approach known as "radical embodiment," and diverges from the abiding tendency of mainstream philosophy of mind and cognitive science to take the unity and homogeneity of the cognitive domain for granted—an assumption that is consistent with the classical interpretation of Cartesian dualism, which these disciplines nonetheless claim to reject.

We then go on to defend the thesis of mental diversity in opposition to the bad Cartesianism that cognitive science and philosophy of mind decline to acknowledge as their own. We argue that scientific developments over the past forty years have fatally undermined the undeclared epistemic imperialism of the theory of extended mind, which assumes the existence of a homogeneous cognitive domain. This assumption, a cornerstone of philosophy of mind and cognitive science still today, is inseparable from the privilege it grants to the subjective experience of knowing, which itself is directly inherited from the methodological solipsism pioneered by Descartes. The assumption of a heterogeneous cognitive domain, by contrast, strips this privilege of all justification. In this respect, it carries on the Copernican revolution of the mind that Kant sought to bring about.

In Chapter 3 we begin with a question that follows on immediately from the foregoing: what, if anything, distinguishes the cognitive system that human beings embody from other cognitive systems? Cognitive science and philosophy of mind refuse to admit the existence of any such special quality with respect to our cognitive abilities, while at the same time, paradoxically, conferring upon human beings an altogether extravagant status as paradigmatic epistemic agents. The explanation that Descartes gave for what he took to be our singular powers is well known: the mind. In his view, the mind constitutes an absolute difference, one that sets us wholly apart from all other living beings, from all other cognitive systems. A careful analysis of the conceptual structure of Cartesian methodological solipsism shows that there is nothing at all absolute about this difference, however. The difference is real, but it is relative, not intrinsic. It arises from the relations between individual people, because the distinctiveness of the cognitive system we embody— what Descartes called "mind"—is indissociable from the fact that we are extraordinarily social beings. The environment from which our characteristic intellectual capacities emerge is a fundamentally social environment.

All of this brings us back to the topic of substitutes and social robotics. If it is true that the human mind is essentially social, then the attempt to create artificial social agents unavoidably amounts to an empirical and experimental inquiry into the nature of the human mind. Social robotics maintains that the fundamental condition artificial agents must satisfy in order to be social agents is the capacity to interact on an affective level with human partners. Emotions are therefore considered to occupy a central place in human sociability. Research in social robotics, in agreement with many recent developments in cognitive psychology and philosophy of mind, recognizes that emotions, far from being opposed to human rationality, are crucial in enabling individuals to develop suitable cognitive strategies. Nevertheless—and in this respect, too, it agrees with the prevailing opinion in cognitive psychology and philosophy of mind—social robotics considers emotions themselves to be internal, private phenomena whose socially interactive dimension is inessential, something added on to emotions, as it were, only afterward. In effect, then, it has taken over from the methodological solipsism common to both cognitive psychology and philosophy of mind, and adapted to the domain of emotions what might be called the "privilege of the first person."

As a consequence of this commitment, current research in social robotics is divided between various approaches that seek to implement the external, public dimension of emotions, and others that aim at giving robots emotions that are conceived as internal, private physiological and psychological phenomena.[13] The contrast between external and internal perspectives is further strengthened by a presumptive distinction between human (and animal) emotions, which are seen as "true," because internal and private, and the affective manifestations of robots, which are regarded as purely external and therefore misleading—benignly misleading, perhaps, but misleading nonetheless, it is held, since there is no corresponding inner state to which they refer. It is this absence of an inner state that the internal approach seeks to remedy. The implications are all the more important as the distinction between the genuine emotions of human agents and the counterfeit affective manifestations of artificial agents is at the center of many ethical debates concerning the use of robots in caring for the elderly, the sick or disabled, and children with special needs. Indeed, the dichotomy between true emotions, supposed to be uniquely the property of human beings (at least in their most fully expressed form, by contrast with the emotions felt by animals), and the false pretenses to which robots are condemned wholly dominates thinking about the dangers inherent in affective relations between artificial agents and human beings.[14]

And yet, as we go on to show later in Chapter 3, the actual practice of social robotics, notwithstanding the separation between inward and outward aspects of emotion that it insists upon as a matter of principle, is founded entirely on the relationship between these two aspects. Furthermore, research in social robotics relies on precisely the fact that the characteristic phenomena of each dimension are mutually self-reinforcing. And so while as a theoretical matter it is the division, on the one hand, between the inward and outward aspects of emotions and, on the other, between true emotions and false affective expressions that guides research, as a practical matter things could not be more different. The reality is that it is impossible to separate these two aspects of our emotional life.

In Chapter 4 we go a step further and call upon social robotics to recognize that its actual practice contradicts the methodological principles on which it claims to be based. Our challenge to the longstanding habit of divorcing the internal and external aspects of emotions receives additional support from two sources: first, from recent advances in the neurosciences, and, second, from the radical approach to embodied cognition in cognitive science. Moreover, as one of us has argued, the dynamic of affective exchange to which the expression of emotion gives rise can be seen as a mechanism for coordinating and reciprocally determining the intention to act on both sides.[15] In this view, what we commonly call "emotions" are salient moments corresponding to equilibrium points of a continuous coordination mechanism. It follows that emotions are the joint result of interaction between two or several agents, not private events produced by the agent's relationship to himself; and that, insofar as they are salient moments, they are public rather than private.

With respect to affective coordination, the relevant question is not whether social robotics ought to try to design robots that have "true" emotions, identified with internal states that are similar (or at least sufficiently similar) to the ones that are supposed to accompany emotions in human beings, or whether, to the contrary, it ought to be content with designing artificial agents that, by pretending to have emotions, exploit our naive anthropomorphism. The relevant question is whether robots can take part in a dynamic process of interaction that, on the basis of affective expression, reciprocally determines the mutual intention to act; that is, a dynamic process of interaction that determines how robots are disposed toward their human partners and how these partners are disposed toward the robots they interact with. We go on to look at three representatives of the current generation—Geminoid, Paro, and KASPAR—and, while recognizing the undoubted successes of their creators, point out the limits to reciprocal coordination that all such robots face.

It is clear, too, that this novel conception of affect cannot help but transform the way in which we think about the ethical problems raised by the behavior of mechanical substitutes in relation to human beings, and vice versa. These problems can no longer be reduced either to a suspicion that the emotions displayed by robots deceive us because of our innate credulousness, or to a concern over the dangers of becoming emotionally involved with artificial agents whose empathy is nothing more than a pretense. The artificial empathy of robots capable of participating in a coordination mechanism is in no way an ersatz empathy, an inferior and misleading imitation of genuine empathy. It holds out the prospect of coevolution with a new kind of social partner, artificial agents that can become a part of the fabric of human relationships in something like the way (while allowing for inevitable differences) that we interact with animals, whether they are pets or animals that have been our neighbors for centuries without ever having been domesticated.[16]

In Chapter 5, the final chapter, we turn to this renewed interest in the ethics of robotics research and to the political issues it raises, taking as our point of departure the development of autonomous weapons and other military robots. This may seem surprising, since military robots are not usually considered to be social agents. Nevertheless, there are good reasons for doing so. Robot ethics not only makes no distinction between substitutes and various types of autonomous agents, it makes no distinction, from the moral point of view, between autonomous agents and autonomous weapons. It seems to us important to show how far robot ethics today is dominated by a particular way of thinking about artificial agents and the role that they may be expected to play in the years to come. Furthermore, the military use of robots is one of the most contested topics in robot ethics, and the morality of their military use is one of the most closely analyzed, not least because it raises immediate and urgent problems that are in no way merely speculative. Here, however, ethicists commonly maintain that one of the great advantages presented by autonomous artificial agents is precisely that they are not autonomous in either a moral or a social sense and that they have no emotions. This position echoes a phenomenon that we mentioned earlier: the rejection of artificial agents that are truly autonomous—autonomous, that is, in the way we are.

We are therefore entitled to wonder exactly what the term "ethics" means in robot ethics, ethics of robotics, and machine ethics. All of these share with the ethics of military robots the same strategy, of ensuring that an artificial agent's behavior will be "ethical" by depriving it of the ability to act other than it is programmed to do. In addition to the question of liberty itself, which plays a central role in moral philosophy, there is a fundamental political issue that must be confronted. It has to do with the difference between two types of autonomous agents and cognitive systems, that is, between social agents, natural and artificial alike, for whom ethical and political questions *are posed,* and artificial agents, which *pose* ethical and political problems for us as human beings.

This latter class of agents, which exist as invisible algorithms distributed within complex systems, decide by themselves what they will do. But here it is not so much a matter of our deciding whether these autonomous agents ought to be allowed to decide by themselves, for example, whether to use lethal force against a potential enemy or to disallow your credit card charge. Because these autonomous agents are in fact cognitive clones, they concentrate in the hands of a few—those who invent them and those who command them—the privilege of making the decisions that structure and transform our common existence. These autonomous agents have never known ethical dilemmas, and they never will; they present *us* with problems that are essentially political in nature, not simply ethical. This is the real issue.

Contrary to what is usually supposed, artificial social actors are likely to enlarge, rather than to diminish, the variety of actions that require us to make decisions. This is why the ethics of our relations with artificial agents cannot be given a general answer in advance. An answer will have to be patiently worked out as more is learned about the nature of human-robot interaction. We give the name "synthetic ethics" to the examination of the moral issues raised by the coevolution of human and artificial actors. This approach is at once ethical and epistemic. It lays emphasis on the need for prudence and wisdom in the sense that the ancients understood these qualities, which is to say as referring to a moral sensibility, rather than to systems of rules and moral principles, as the arguments advanced by robot ethicists typically do today. We call this kind of ethics "synthetic" because, in the context of coevolution, knowing is inseparable from doing; because discovering new moral rules is inseparable from creating new social agents, from inventing new machines—none of which have ever been seen before.

CHAPTER 1

The Substitute

[O]perative is not a synonym of practical; the operation of technology is not arbitrary, turned this way or that as the subject pleases, depending on its momentary usefulness; technology is pure operation, bringing into play the true laws of natural reality; the artificial is induced nature, not something falsely natural or something human mistaken for natural.

GILBERT SIMONDON

If forecasts of scientific and technological progress, for example with regard to the course of development in robotics, generally prove to be so shortsighted, so disappointing, and so often refuted by subsequent events, it is, among other reasons, because they take it for granted that the world is something already given, something out there that science discovers and technology then transforms. And so they imagine that it must be possible, by looking at the development of existing technologies, to make out the features of a yet unknown future. But the transformation of the world is not something that happens later, once science has made the necessary discoveries; to the contrary, it is part and parcel of the act of discovery itself. Nevertheless, this transformation is in no way random or arbitrary. As Gilbert Simondon says, technology brings into play the true laws of natural reality. It is therefore not by chance that the fields in which social robotics has so far penetrated most deeply should be health services and medical care, on the one hand, and, on the other, the military sector.[1] We are now poised to entrust to substitutes and other artificial agents two fundamental human activities associated with cooperation and conflict, respectively: helping those human beings for whom we feel a sense of responsibility, and killing others. Beyond the evident ethical problems these developments raise, they illustrate the close relations that have always obtained between conflict and cooperation. There is no technological determinism in the development of such artificial creatures. Nor are science and technology radically changing human society. The artificial creatures we are already making now, and the ones we plan to make in the future, reflect basic aspects of human sociability.

The Uncanny Valley

More than forty years ago, the Japanese roboticist Masahiro Mori advanced a conjecture known as the "uncanny valley" effect.[2] Although it has never been proved, his hypothesis continues still today to guide researchers in their efforts to understand and improve social interactions between robots and human beings. Mori postulated that the more robots resemble humans, the easier and more comfortable we find it to be with them—but only up to a point. Beyond a certain threshold, the likeness proves to be too close and yet not close enough. The rising curve in Figure 1 represents the growing sense of ease we feel in the company of robots the more they resemble us. Suddenly the line turns downward: now we avoid approaching these strange and disturbing androids, which are at once too much like us and yet not enough like us. This is the uncanny valley. It opens up when the still imperfect resemblance between humans and robots becomes too nearly perfect, and it closes once robots have become indiscernible from humans. The curve then resumes its upward course: we gladly interact once more with artificial creatures, which now look more and more like the Buddha than the ordinary person.

The received interpretation of the uncanny valley effect is that very small differences, the visible and tangible discrepancies resulting from minutely inexact duplication, make interaction with robots that

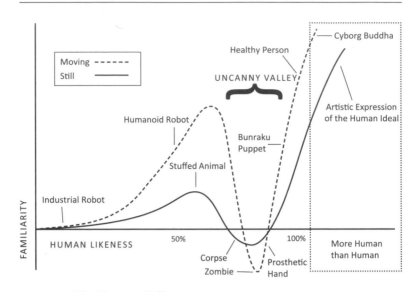

FIGURE 1. The Uncanny Valley

almost but not quite resemble us unbearable. We find these tiny differences disconcerting, sometimes shocking or even repellent. This is what Mori has in mind when he takes as an example the experience of touching the hand of an android that until then seemed human to us. We suddenly have the impression of shaking the cold and rigid hand of a living corpse. The terrifying surprise paralyzes us. Here we have reached the bottom of the uncanny valley.

Some roboticists who have carefully studied this effect seek to reduce the differences in appearance between man and machine as far as possible, with the aim of making a robot that exactly resembles a human being.[3] In doing this, they hope to be able to determine the precise point at which the sense of familiarity gives way to a feeling of danger. One idea they explore is that uneasiness is a consequence of too great a physical resemblance. This suggests that what frightens us, what we find threatening when faced with extreme similarity, cannot be anything that distinguishes us from robots and makes them strangers to us, for we encounter nothing of the kind, or at least vanishingly little. What scares us is the very *absence* of such distinguishing marks—with the result that we have a hard time reassuring ourselves that, no matter how similar they may seem, robots are different, they are not really like us, they are not really human beings.

Why should this be so? First, because there is nothing simple or straightforward about interacting with other human beings, least of all when they are unknown to us or when social conventions break down. In that case, anxiety and fear are never far away. Contrary to what the usual interpretation of the uncanny valley assumes, contact between people who do not know each other is often abrupt, and almost always unsettling, when it takes place with no reference to a shared sense of appropriate behavior. We need to have a code of conduct, a common language and set of rituals, if we are to be able to interact in a relaxed way and feel comfortable with one another. Culture itself may well be an invention, continually modified, that allows a large number of individuals to live together more or less easily.[4] Second, though this is only another aspect of the same difficulty, one might suppose that the real reason we seek to make machines that are *like* us is precisely that they will not *be* us. We want them to be different, in certain respects better and more efficient than we are, but in others less good; or at least we want them to behave differently than we do. We want robots to be mere machines that we can control at any moment, that will not be as rebellious and unpredictable as other people—which is to say ourselves, since each of us is an other for everyone else.

What we find most unsettling, then, are not the differences between robots and ourselves, but the idea that an artificial agent might begin to act just like a human being, with all that this entails in the way of uncertainty, unpredictability, and danger. The robot, the artificial agent that we have built, would then become altogether as unknown to us as our fellow human beings are.[5] In Karel Čapek's play, which made the word "robot" part of everyday speech almost a century ago, robots are mechanical (or, rather, biosynthetic) slaves that end up rebelling when they become too much like human beings. From the very beginning, robots were imagined to be human beings who are not quite human beings: the danger they present is that they may actually become human. A robot can do everything we can do, but it is not one of us. And it is this difference, for better or for worse—or, more precisely, for better *and* for worse—that gives the robot its true worth.

That the problem may be an excess of resemblance, rather than a lack of it, was suggested by the figure that Mori himself drew to illustrate the uncanny valley conjecture (see Figure 1). Surprisingly, this figure does not stop at what, at first sight, may seem to be the objective to be reached, a robot that fully (100 percent) resembles a healthy person. The curve that represents the feeling of affinity, or familiarity, and ease of interaction continues to rise beyond this point, up and into a region identified as "more human than human" that includes "artistic expression of the human ideal" and, at a still higher level, a "cyborg Buddha." The continuation of the ascending curve suggests that perfect resemblance is not in fact ideal, and that interaction will be more comfortable once robots cease to perfectly resemble us. There is therefore on one side of the uncanny valley a negative difference, as it were, where the robot does not yet sufficiently resemble us. The ascending curve indicates that, as the robot resembles us more and more, the gradual elimination of this negative difference makes our relations with robots easier and more agreeable—up until the catastrophic moment when the resemblance becomes too great. We leave the valley when the robot completely resembles a healthy person. The curve resumes its upward path on the far side of the valley, indicating that our relations with robots have once more become progressively easier, only now they display a positive difference. Here the difference measures the distance separating us from a robot that is superior to us, a robot that, like the Buddha, is more perfect than we are. The sustained rising course beyond perfect resemblance as this positive difference increases suggests that too great a resemblance—too close an approximation of identity, of indiscernibility—is indeed the source of the problem.

If Mori does not comment on this mirror symmetry, it is because the curve's continuing ascent on the far side of the valley points to the potential superiority of artificial creatures. Once again they resemble us, without being exactly the same: though in certain respects they are inferior to us, in other respects they are superior, or at least express an ideal of what we may aspire to be, for we seek in the mechanical beings that we create a perfection that eludes us as human beings. There is nothing new about this. Philippe Pinel, in section III of the *Traité médico-philosophique sur l'aliénation mentale, ou La Manie* (1801), entitled "Of Malconformation of the Skulls of Maniacs and Idiots," sought to determine "how far the best proportions of the head are to be considered as external indications of the excellence of the intellectual faculties." As a basis for comparing the skulls of the insane, he chose "the beautiful proportions of the head of the [Pythian] Apollo."[6] Abnormal deviations, in other words, are identified and measured with reference to a head modeled on an artistic ideal that "unites the best proportions and the most harmonious lines that are possible to be met with in the most perfect configurations of life."[7] This norm is evidently not statistical; it is an ideal norm. Similarly, Mori conjectures that we will feel more at home with an artificial being that expresses a human artistic ideal than with one of our fellow human beings—with a "cyborg Buddha" that is "more human than human." For Mori, no less than Pinel, the artificial most fully represents an ideal in relation to which we all fall short. Artifice, in this case a machine we ourselves have created, has the power, at least in principle, to disclose the secret of human potential.

The idea that artificial beings might be superior to us, and that they are messengers from or inhabitants of a more perfect region of existence to which they can give us access, is in fact a very ancient one.[8] Since the earliest times, statues that speak or that move have been accepted as divine envoys or as incarnations of gods themselves. One must not suppose that we are dealing here only with primitive mythology or with a ruse practiced by priests and other clever persons to fool the naive. In Rome, as John Cohen reminds us, the mechanism that caused the statues to move and the priests who operated it were sometimes fully visible. The "deception" was therefore carried out in plain sight, known to all.[9] It was not, in other words, a deception at all, or else it was one of a very different kind from what we usually understand this word to mean, since it depended mainly on a suspension of disbelief. It relied not on deceit or misrepresentation, but on a readiness to believe, to play along.

Mori, in his way, agrees with the idea of a suspension of disbelief. Still today, almost fifty years later, artificial creatures populating the land that begins on the far side of the uncanny valley come under the heading of science fiction. For the moment, at least, the world that he imagines beyond the uncanny valley does not exist. It may never exist. And yet it belongs to a very old realm of the human imagination that seeks in our own creations the ideal of what we might be, and that fears, above all, discovering in them what we really are.

More than half a century ago, the philosopher Günther Anders gave the name "Promethean shame" to the shame we feel when faced with the perfection of our own creations, the shame that "*takes hold of man before the humiliating quality of the things that he himself has made.*"[10] The uncanny valley, we believe, shows that the superiority human beings are prepared to concede to machines and artificial creatures of their own devising, whether it is manifested in the form of shame or of admiration, is inseparable from the difficulty they have in living together—a difficulty that is revealed by the uneasiness humans feel when robots begin to resemble them too closely.

Robots as Substitutes

Robots are in fact very different from one another. They are not all created equal in our eyes, nor do they all resemble us in the same way: none resembles us perfectly, and each one resembles us more in certain respects than in others. Some are used in cooking, some are built into airplanes, some are employed in factories, some are deployed as autonomous weapons; still others are meant to function as *substitutes,* as we propose calling them. The term "robot" does not encompass a natural kind. It does not refer to a homogeneous group, but to a polythetic class of objects that are related to one another by rather loose family resemblances and share no common property that is uniquely theirs. Within this rich and diverse community, substitutes are the only ones that make us feel uneasy, that captivate us and disturb us in something like the way our fellow human beings do. And like our fellow human beings, they may, by their very nature, be either rivals or allies; inevitably, they turn out to be both.

We normally think of a substitute—a substitute teacher, for example, or a deputy prosecutor—as someone who replaces someone else *without taking his place;* that is, without taking his office away from him. A substitute thus makes it possible for the person he replaces to do more than could be done without his aid; to be (in effect) somewhere that (in fact) he is not, for example, or to handle several matters at the same time. The substitute must have at least some of the qualities and aptitudes of the person whose role he plays, without the role-playing in this case being only a game, since the substitute partakes in the authority of the person whom he replaces, so that his decisions and his judgments have the same force. This sharing of the authority possessed by the person being substituted for is essential if the substitute is to be able to act on his behalf. It is sometimes the case that the substitute, after a more or less long period of time, inherits the responsibility of the person for whom he substitutes, but not always. It may be that he proves to be incompetent. In most cases, however, it is neither required nor expected of the substitute that he will take the place of the person on whose behalf he is acting. The purpose of the substitute is to stand in for another person while that person is absent, or momentarily unable to perform his duties, or incapable of going to a certain place himself. A substitute is an authorized surrogate, like a minister plenipotentiary, or an honorary consul who plays the role of someone else, but only in part; that is, only in certain cases, in certain respects, and at certain moments. It is by virtue of just this quality that he is a substitute.

The aim of creating artificial surrogates, which is to say robots capable of substituting for us in the performance of certain tasks, but without taking our place, is central to social robotics—not in the sense that the majority of roboticists working in this field would recognize it as characterizing their activity, but in the sense that it is often implicit in what they do. To create an artificial nurse's aide, a robot companion for the lonely, or a mechanical nanny that takes care of children while their parents are away, indisputably amounts to introducing into our ordinary social relations artificial social agents that substitute for human (or animal) agents, but without taking the place of these agents. Artificial nurse's aides, robot companions, and mechanical nannies do not render our relationship to nurses, parents, and friends obsolete or useless, nor are they intended to do away with them. They are meant to stand in for them at certain times and under certain circumstances.

This is not true of all robots with social functions. It is not true of Reysone, for example, the robotic bed that converts into an electric armchair to allow persons of limited mobility to move more easily and with greater independence from the recumbent to the seated position. Reysone increases the personal autonomy of patients by eliminating their reliance on a nurse to sit up, but it is in no way a social agent. It is not a substitute. Similarly, an industrial robot on an assembly line takes the place of human workers. It does not simply occupy the place they once occupied in the factory; it transforms their job and may even cause their job to be abolished. Indeed, this is precisely the economic point of the capital investment that the robot represents in relation to labor costs.[11]

The matter is quite different with substitutes. Artificial receptionists, nurse's aides, and animal companions are surrogates that occupy the place of someone momentarily. They perform his or her duties, but they do not replace him or her permanently. They do not take the person's place. Our interest, then, is not in the entire class of artificial agents that are commonly called social robots, but only in those that are, or that aim to be, substitutes. Most, if not all, of them succeed in this only partially and imperfectly. The ambition to develop more perfect substitutes drives current research in a number of areas of social robotics, where our technological ingenuity is now beginning to realize at least some of the dreams of science fiction.

Substitutes constitute a particular type of technological object. They are in several respects similar but not identical to slaves according to Aristotle (*Politics* 1.4, 1253^{b}23–39), animated instruments— "living tools." Substitutes occupy our place at certain moments, occasionally perform our duties, but without taking them over from us. They make it possible for us to do what we could not do without them; like slaves, it is them who do that which we could not do without them. Substitutes are robots, however, not human beings. Nevertheless, like many slaves through the ages, and unlike mere tools, substitutes are granted an authority that they would not otherwise possess. Insofar as being a substitute consists in having a certain legal status, creating a substitute is something that, if not easily done, can at least in principle be done openly, with full and proper authorization. Insofar as being a substitute consists in exercising legal authority, it is not an *object* of a particular type; it is essentially a role, an activity that exists by virtue of an act that, in explicitly assigning the substitute a particular office or function, specifies the limits of his authority. But how is a substitute, a surrogate, to be invested with lawful authority when it is not a person, but a thing? How are we to fabricate something that may properly be described as a substitute—that is, an inanimate object that is capable of exercising authority?

A Robot that Serves No Purpose

"I'm trying to make a robot that doesn't serve any purpose," Takanori Shibata replied when asked about his ongoing research project that became Paro, a robotic baby harp seal. Though slightly misleading, his reply nevertheless makes the essential point: Paro is intended to be a substitute animal companion in hospitals and retirement homes, and yet it has no specific function.[12] In other words, it must be able to do any number of things: entertain, comfort, reassure. . . . The list goes on indefinitely. More precisely, it is undefined, since to give company, unlike washing the dishes or vacuuming, is not something whose various aspects can be exhaustively enumerated in advance or that can be sufficiently restricted by external constraints so as to rule out the unforeseen. Even when it serves a particular purpose, a social agent, unlike an automatic security gate or a self-operating lawn mower, must be able to disengage from that task. It must be able to adapt to a wide range of different situations, to stop what it is doing at any given moment and turn its attention to another task.[13]

Social creatures do not serve any one particular purpose. They need to be able to perform a variety of duties at different times. By definition, a social robot that functions as a substitute cannot be restricted to a single role or function; otherwise it would not be a social creature. More generally, and more profoundly, sociability has no particular end other than itself. It is the necessary condition for the emergence of any end or purpose within a material culture, which itself limits the range of possible ends. Sociability is a fundamental part of what it means to be human. All the aims and all the purposes that we can have are determined with reference to it. These aims and purposes are ones that we give to ourselves as social beings or that can be imposed on us from outside. While we can be reduced by circumstances or otherwise forced to go on repeating the same action over and over, the ability to transcend a role or a function in which other people try to confine us, or to which the order of things tends to restrict us, is inherent in our condition as social beings. Human sociability supposes an indefinite capacity for the reciprocal coordination of individual behaviors.

It is precisely this capacity for adaptation, this ability at any moment to *reconfigure* their relations to one another that confers on slaves, servants, and human workers their great superiority (and that likewise constitutes their principal shortcoming) by comparison with machines.[14] Being able to change the nature and the form of socially coordinated behavior, while it permits human beings to adapt themselves to a much larger range of situations than any of the machines that we presently know how to build, also makes attempts to control them more difficult and, where successful, harder to sustain. It is furthermore one of the major reasons why we are in many cases tempted to dispense with our fellow human beings, to replace them with tools, instruments, or technological systems that are not liable to sudden and unforeseeable changes in behavior. Every hierarchical organization seeks to circumscribe this indefinite capacity for coordination, or even circumvent it, by devising limited and fixed tasks and, if possible, entrusting them to machines of one sort or another that are dedicated to a single purpose. This habit is the source of all those hybrid "assemblages" that Bruno Latour is so fond of.[15] But while the specialization and specification of tasks and functions, which make it possible to exploit technological objects in order to constrain and limit the human capacity for coordination, are an essential dimension of human sociability,[16] they are only one dimension. Unlike many other technological objects, the subset of social robots that we call substitutes do not aim at restricting the human capacity for coordination, but at imitating it, at becoming part of it.

At the present moment, there exists no artificial agent capable of making dinner, mending a pair of pants, running a vacuum cleaner, taking children to school, gardening, hunting peccaries in the jungle, trimming a Christmas tree, playing the flute, and telephoning home to say that he will be late. The problem is not simply technical. Many expert systems are able to perform one or another of these tasks, such as telephoning, playing the flute, or vacuuming. Other such systems, capable of gardening, taking children to school, mending a pair of pants, making dinner, or hunting peccaries, seem well within our reach and indeed already exist in the form of prototypes.

And yet beyond the currently insurmountable technical obstacles to creating so versatile an artificial agent, building an artificial agent capable of performing all of these tasks, or at least several of them, presents difficulties of an entirely different order. To invent a device capable of carrying out one or another of these functions is essentially an engineering problem, but to create an artificial agent that can move from one to the other spontaneously and naturally—not only harmoniously, in the narrow sense of moving from one to the other in a fluid manner—poses difficulties that are social and relational in nature.

The list of tasks that a social being may be called upon to carry out is by definition open-ended, being both infinite and indeterminate. It is impossible to know in advance the next item that will appear on the list, the next move that will need to be made. A social robot in the sense that we intend must therefore be able to adapt to new and unforeseeable situations. To be sure, by "new situations" we do not mean any situations whatever. The logical space in which they can appear is not unlimited, nor in any way equivalent to the vague and poorly defined domain of things that human beings can do. If we are in a position to fabricate artificial agents that can play social roles, it is, among other reasons, because it is possible in advance to rule out as highly improbable a vast range of alternatives—and this because there exist social roles that are well-defined. A substitute is first of all a robot that is capable of carrying out certain tasks, as a receptionist, for example, or a nurse's aide. Yet it must be able to react in real time, to adapt, to coordinate its actions with those of its human partners. While it is possible to imagine scenarios, to define functions, to control the environment in such a way that the behavior of human and artificial agents will be both restricted and predictable, it is nonetheless impossible to eliminate entirely the dimension of reciprocal coordination. Artificial social agents are different from other technological objects in that they must necessarily exhibit this capacity for adaptation and coordination, if only in some limited fashion. And this capacity requires that the robot be able to interrupt the performance of a particular task or function. Artificial sociability will have been fully realized, as Shibata well understood, when it will be possible to create a robot *that serves no purpose, because it can serve any purpose.*

Being Present

A second characteristic of social robots that distinguishes them from many other modern technological objects is their ability to be present. We noted earlier that a substitute makes up for the absence of the person on whose behalf he acts. To be present in this sense constitutes an action. Many modern technological objects, by contrast, are not meant to be noticed while performing the role for which they are designed. They are meant instead to be more or less absent. An object's disappearance within its function, as it were, is particularly remarkable in the case of information and communications technologies.[17] The innumerable screens that we look at every day—the screens of our televisions, our computers, our telephones—remain for the most part in the background, obscured by the images of something or somewhere else that are projected on them. The physical presence of a computer or a smartphone really makes itself felt only when it breaks down or when our relationship with it goes beyond merely operating it—for example, when we move a laptop to another room or another office, when we use a phone as a paperweight, or simply when we have put the phone down somewhere and try to locate it. When they are functioning normally, however, these devices, though they are omnipresent in our daily lives, make themselves effectively invisible as physical objects. So long as they do what they are designed to do, we tend to see through them. We seldom pay any attention to them at all, only to what they do or what they allow us to do.

The fierce rivalry between leading manufacturers to create the lightest technological object—or the thinnest, the smallest, the least noticeable object, or an object that, outwardly at least, is indistinguishable from one of the portable devices that we commonly wear on our person, such as an intelligent wristwatch or eyeglasses with integrated displays—shows that this characteristic of information and communications technologies also implies a certain business strategy. The most successful object commercially, it is believed, will be the one that comes closest to achieving the dual objective of technical high performance and physical unobtrusiveness. This line of thought reaches its logical conclusion with the aim of creating cyborgs: the computer chip, once inserted beneath the skin and made part of the user's body, now completely disappears as an object; there remains only what it allows us to do.[18]

There is a second form of absence associated with these technologies, inseparable from the first. This time it concerns the user, not the object. Computers, tablets, smartphones, and the like open onto a different space, an alternative world, parallel to the one we inhabit and yet distinct from it—the so-called virtual world of the Web. This relation space has no distance. It is filled with communications that are indifferent to remoteness or proximity in physical space and yet still depend on the localization of agents in relation to the network that supports such communications. Whoever has access to this space is no longer present in the same manner in the physical space of human sociability that he nonetheless continues to occupy. He is at once here, in front of a screen with his fingers on a keyboard, and somewhere else. But absence in this sense from physical space, where inevitably we go on existing, is not the same as the absence of someone who is absorbed in the book he is reading. The somewhere else in this case is not simply mental, as when one reads, or dreams, or is distracted. Nor is it merely virtual. In this elsewhere it is in fact possible for us to act and interact with others who are not immediately present, here and now. It corresponds to a different space from physical space, a parallel space in which we can also be present. The reason it is not merely virtual is that actions performed within this space can have repercussions in the three-dimensional physical space in which we live—indeed, they are liable to transform the world. This space therefore authorizes a kind of presence that is no less real for being both disembodied and delocalized, for it may nonetheless influence, may spill over, as it were, into the world of shared physical presence. In this case, our "absence" from shared physical space is inseparable from another kind of presence, in an elsewhere that in principle is at once nowhere and anywhere.

Active presence in shared physical social space is, to the contrary, a fundamental characteristic of substitutes, a necessary condition of their being able to play their role. A robot is a three-dimensional object in physical space. This is precisely what distinguishes it as an artificial agent from virtual agents that have only a numerical and digital existence projected onto a screen.[19] Social robots are solids. They do not distance us from physical space, and they do not invite us to remove ourselves from it. On the contrary, they encounter us within it, and we encounter them there as well. They do not disappear in the course of being used, but instead affirm their material presence by acting. An artificial nurse's aide, a mechanical adjunct to therapeutic treatment, and a substitute animal companion all do what they do by intervening in shared physical space. These robots do not cultivate evanescence; they do not simulate immateriality. They are among us in an altogether different manner than the many portals leading to the deceptively virtual universe of the Web. They have a physical social presence. They make themselves apparent. Their ability to act in shared physical space is one of the reasons why we make robots that can do what disembodied virtual agents are incapable of doing.

The presence displayed by substitutes is also different from the *telepresence* that certain robots make possible. Like substitutes, these robots are embodied agents and can, at least in principle, move and act in physical space. Like substitutes, too, they are endowed with a social presence, but it is not really theirs; it belongs to the human agents who control them from somewhere more or less far away in physical space. Substitutes, by contrast, substitute for people in whose place they act. They are autonomous representatives, not mere representations. This makes them independent social interlocutors, rather than three-dimensional images that lend a body to the action (essentially a verbal action, by which words are transmitted as over a telephone line) of someone who is in another location. Substitutes act by themselves. The social presence they embody is their own; they are not limited to representing someone who is absent.

Authority

A third fundamental characteristic of substitutes is that they are mechanical devices endowed with an authority that they can exercise. Many devices are capable of *being authoritative:* the tape measure we use to determine the dimensions of a room, a thermometer, an altimeter, a depth probe—indeed, any measuring instrument that furnishes authoritative results, unless we have reason to believe that it is functioning incorrectly or that it has been miscalibrated. Strictly speaking, however, such objects exercise no authority over us. We are under no obligation to accept the information they give us, even if it would often be unreasonable to disregard it. The failure to take into account the result or the measurement an instrument gives is at the very most an error, a mistake; it is in no way a transgression, because it does not contravene any authority.

Are there other technological devices that can be said to possess authority in a strict sense? Is there any reason not to say, for example, that the automated information system that approves or rejects a credit card charge on the basis of your past transactions and your travel and consumption habits has the authority to authorize or disallow payment? This way of talking is, at a minimum, deeply misleading. It overlooks the fact that such a system eliminates any possible relationship of authority and replaces it with a fait accompli or, more precisely, makes acting otherwise impossible. It replaces authority as a type of social relation with another means of obtaining the desired result: giving you access to your funds and protecting against fraud. But it neither expresses nor embodies any authority, whether of a bank or any other issuer of credit.

Whoever has the authority to make a decision, to allow toxic waste material to be stored in a particular place, for example, is responsible for the decision made. He or she must be able to explain the reasons for the decision and will be subject to legal action and penalties in the event that the decision proves to have been disastrous, lacking in justification, or made without the requisite authority. In the case of credit card authorization, a verification system accepts or refuses your demand for advance payment. That is all there is to it: no one is responsible. A previously determined algorithm carries out a statistical inference whose result is what might very loosely be called the decision—or, better, the *decree*—of a computer. While speed and efficiency are probably the main advantages obtained by automating the decision procedure in this case, it is probable too that the disappearance of anyone to whom its responsibility might be assigned constitutes a nonnegligible marginal benefit.[20] This disappearance of responsibility is a sign that an authority relation has been transformed into a power relation, even in cases in which power is not exercised in the form of physical coercion.

Authority exists only to the extent that it is recognized. Here, however, it is impossible to identify anyone as being responsible or in any way authorized. The decision procedure is obscure and incomprehensible; indeed, most people are not even aware of the existence of an algorithm that accepts or refuses demands for payment. The transformation of an authority relation into a power relation does not arise simply from the fact that if your demand is refused you then find yourself without recourse or remedy (at least not immediately), but also from the fact that the procedure dispenses with your consent to the decision by which you now find yourself bound. Every authority relation, after all, is to some extent a political relation.

A teacher has authority in his class if, when he asks unruly students either to cease being disruptive or leave, they cease or they leave. In either case, they recognize the teacher's authority and accept the choice with which they have been presented. Otherwise, if the teacher has to use force to expel them, he does not have authority over them. Even if the threat of force is very often a last option, authority exists exactly to the extent that it is not necessary to resort to it. This is why resorting to force in an attempt to reestablish an authority that has been undermined frequently fails. The automation of many daily activities replaces authority relations with power relations by imposing judgments whose decision criteria are unknown, in most cases without there being any immediate possibility of challenging a judgment, without it being possible even to find out to what person or office such an appeal might be addressed. Technological processes of this sort exclude any kind of relationship to a properly constituted authority. They replace recourse to authority with an accomplished fact.[21]

The simplification of social relations, which limits and constrains them while at the same time transforming the environment—a social environment—in which these relations obtain, is a luxury that substitutes cannot afford. To do their job, whether as a nurse's aide, a nanny, a kindergarten teacher's assistant, or a receptionist, they must be able to establish authority relations and to maintain the authority with which they are invested as substitutes. A robot presents an advantage in looking after children, compared to a camera system, only if it is capable of intervening in their activities, in diverting them from some and steering them toward others. It must be able to do this without resorting to force, all the while being capable of defending itself, which is to say of resisting attempts to disarm it, evade it, or nullify its effectiveness in some other way. A robotic social agent must be an interlocutor that manages to assert itself to one degree or another and to make itself respected. In this sense, the behavior of substitute social robots is utterly opposed to the silent, anonymous, and invisible behavior of automated systems that deprive us of the possibility of acting otherwise than we do.

Technological Objects and Social Actors

Authority, active social presence, and an effectively unlimited capacity for coordinating their behavior with that of human partners are three characteristics that distinguish substitutes from other technological objects. Artificial agents of very different types are nonetheless currently being developed in social robotics laboratories. We have already mentioned robots that keep elderly people company at home or in an assisted-living facility, make sure they take their prescribed medicines, help them move around or take a bath, and are in constant communication with a central operator, human or robotic. They are capable of asking for help in case the people in their care refuse treatment, fall, or are for whatever reason in danger. Such robots protect by restraining. They do this without coercion, simply by making certain behaviors impossible or inaccessible, such as refusing to take medicines or attempting to move or act in a way that is "judged" to be unsafe. Artificial caregivers of this sort are often put to use in so-called smart homes that are capable of monitoring all the movements of their occupants; ideally, rooms can be modified to meet their occupants' needs or to prevent accidents. The settings in which these interactive robots operate are panopticons of variable design, in other words, in which the well-being of the inhabitants is ensured by a kind of restraint that is as gentle and as quiet as it is firm. These robots are social robots, and the controlled environments in which they function come under the head of social robotics, but they are for the most part technological objects of a different kind than substitutes.

Bruno Latour has rather humorously argued, but in all seriousness just the same, that a wide variety of technological objects, often very simple ones—a spring or a hydraulic piston that closes a door automatically, for example, or a gate opener equipped with a photoelectric eye, or even a spring-loaded turnspit—should be regarded as social actors, which he calls "actants," on the same level as human beings. Social actors, he maintains, cannot be divided into human actors on one side, and technological artifacts that are used by human actors and that influence their behavior on the other. If one wishes to understand social action and the place of technologies within societies, one must treat human and nonhuman actants symmetrically: "Students of technology are never faced with people on the one hand and things on the other, they are faced with programs of action, sections of which are endowed to *parts* of humans, while others are entrusted to parts of nonhumans."[22]

The case Latour makes is interesting and, at least to some extent, persuasive. Note, however, that in his preferred example, the automatic door closer, though a nonhuman actor does the work of a human actor, it does not replace him. One may wonder of course if in this case an employee's freedom of action has not already been so reduced by the profit-maximizing demands of modern business that the loss represented by the disappearance of his job is rather small, perhaps even insignificant compared with the advantages in terms of both economics and efficiency of a mechanical actor (the door always closes as long as the mechanism does not break down).[23] Here, what makes it possible to replace a human being by an artificial mechanism is not only the existence of a suitable technological object, but also a prior transformation of social relations. Nonetheless, even where recourse to a nonhuman actant is facilitated by the relative desocialization of the task needing to be accomplished, the nonhuman actor's abilities are extremely limited compared with those of the social agent it replaces. In a hotel, for example, the automatic door system is unable to respond to strange or unusual questions, to object to a guest's outrageous behavior, to come to the aid of a staggering drunk, or to stop children from opening and closing doors for fun.

It is by no means impossible, at least not in principle, to create an artificial agent capable of performing these routine acts of human sociability, rather than settle for the limited program of action of a mechanical door closer. Yet an artificial agent capable of genuinely social behaviors is a very different technological object from a door-closing system; it is very different, too, from an electronic surveillance system that breaks down social behaviors into discrete tasks and distributes them among several human and nonhuman "parts." These differences arise precisely from the capacity for sociability exhibited by a porter or a bellhop or a robot substitute. It rests on the fact that the social agent, human or robot, makes its presence felt by those with whom it interacts; that it can adapt to many situations; that, depending on the needs of the moment, it is able to coordinate its behavior in many different ways with that of its interlocutors; and that it is endowed with at least minimal authority. These are all abilities that a door closer does not have, any more than the spring-loaded turnspit of a rotisserie does. Automated devices of this sort have no social presence; when they function correctly they are, for all intents and purposes, absent. They have a highly limited capacity for adaptation with regard to the task they are designed to carry out—in Latour's terms, the section of the program of action that is assigned to them—and they are wholly devoid of authority. From the technical point of view, creating substitutes is an entirely different proposition.

Autonomy

The three characteristics of robots we have just considered—authority, presence, and indefinite capacity for adaptation and coordination—jointly imply that substitutes enjoy what may be called *social autonomy*. Not all social robots are autonomous technological objects. Some are remotely guided by a human operator. Playing a social role, which is to say functioning as a substitute, nevertheless requires that they display autonomy of a particular kind.

David McFarland, an authority on animal behavior and robots, distinguishes three forms of independence that robots may exhibit, corresponding to three degrees on an increasing scale of autonomy.[24] First, robots must have energy autonomy that constitutes the necessary condition of the second form, motivational autonomy; together, they constitute the necessary condition of the third form, mental autonomy. The first form consists in an artificial agent's ability to meet its energy needs unaided. This is the case, for example, with a robotic lawn mower that automatically plugs itself into an electrical power source when it detects that its batteries are beginning to run low. McFarland holds that a robotic lawn mower also possesses motivational autonomy, to the extent that it determines its own behavior as a function of what it "wants" to do, namely, recharge its batteries or cut the grass. Its motivations, he says, are all under its own control, in the sense that no person is needed to tell it what to do. As for mental autonomy, it can be ascribed only if the lawn mower is the author of its own mental acts. In that case, its mental acts would be under its own control as well. As McFarland puts it, "The mentally autonomous robot would be one that is free from outside control of its mental states."[25] This is not true of a robotic lawn mower, however, for its "mental states" are indeed under outside control. It *wants* to mow the lawn or to recharge its batteries only because it has been programmed to do these things under certain well-defined circumstances.

McFarland's classification of the forms of autonomy presents many difficulties. The example of a robotic lawn mower is revealing of these difficulties because in this case it is its motivational autonomy that makes possible its energy autonomy. It is because it goes by itself to recharge its batteries when they are running low that it is able to meet its energy needs. It is autonomous in terms of energy, in other words, because it takes by itself the necessary measures to ensure that its batteries will never be completely empty. If we had to plug in the lawn mower ourselves, it would not have energy autonomy. Without the ability to recognize its own "internal state" (the degree to which its batteries are charged), to point itself in the right direction and make its way to the recharging station, and to do what is needed itself, it would have no energy autonomy whatever. In this example, at least, one would be hard-pressed to say that we are dealing with two different types of autonomy or even two distinct degrees of autonomy.

Similarly, one may wonder whether, for example, a lamp equipped with a photovoltaic cell that during the day stores energy from sunlight, which it then uses to provide illumination at night, can really be said to be autonomous in terms of energy, for it does nothing. Being bright and being dark are things that happen to it—states, rather than actions, that are wholly determined from outside. A number of years ago now, Humberto Maturana and Francisco Varela put forward a quite different conception of autonomy. In their view, an autonomous system is a system that, in acting upon itself and its environment, is able to preserve its functional integrity, that is, to maintain certain fundamental parameters within a certain range of values.[26] This is what the solar light in our example is unable to do. Up to a certain point, however, though still to a very small extent, the lawn mower is able to do this, by making sure that its battery charge never drops below a certain level. We are therefore not dealing here with two distinct types of autonomy, or with two more or less considerable degrees of autonomy, but with a single capacity, namely, for acting in such a way as to maintain one's power to act.

Autonomy, understood in this way, is not an intrinsic characteristic of a system, but a relational property. It consists in a system's ability to do certain things by itself that will permit it to continue to exist and to function normally. Note that this ability does not imply that action is independent of the setting in which an agent acts. There must be an appropriate environment, furnished with everything needed to ensure that the functional organization of the system will be protected and perpetuated. Thus, the lawn mower can exercise its so-called energy autonomy only in a suitable setting, one that contains a central power supply and a recharging station that it can plug into. It can enjoy this autonomy only if the topography of its environment allows it to have access to the recharging station. The same is true of its so-called motivational autonomy. This form of autonomy is limited to a choice between two options: "wanting" to mow the lawn and "wanting" to recharge. These options are only meaningful in a predetermined and narrowly specified environment where they correspond to real possibilities of action for the robot.

This is why the third form of autonomy identified by McFarland is not merely confused, but actually incoherent. It requires that the agent be independent in relation to the environment in which it physically intervenes, or rather thinks and deliberates, since here it is no longer a question of action; what we are dealing with is an exercise of imagination or of thinking that takes place outside any environment. If, as McFarland maintains, the robotic lawn mower has no mental autonomy, it is because its so-called mental states—the alternatives of cutting the grass and recharging the batteries—are the result of external action by a programmer. The robot is incapable of providing itself with any mental states that are not under foreign control. Yet since the lawn mower's mental states are identified with its motivations, if the argument were valid it would be necessary to conclude that the lawn mower has no motivational autonomy either, for it is no more responsible for its motivations than it is for its mental states. It would be more accurate to say that the lawn mower has just as much mental autonomy as it has motivational autonomy[27] and energy autonomy—for these are not three distinct forms of autonomy. In other words, very little!

There is no such thing as pure or absolute autonomy. Mental autonomy, in the sense in which McFarland understands it, not only does not exist, it is impossible.[28] Autonomy can only be relative and relational, for it involves actions that are only possible within an appropriate environment. Total independence with regard to one's surroundings is an illusion. Autonomy occurs in the context of a range of possible actions that transform one's surroundings. The autonomy of an agent is always relative to a certain setting or milieu, and it gives rise to a history. The options an autonomous agent selects—which is to say the actions a robot or an organism undertakes in order to attain its objectives—inevitably have the consequence of altering the environment in which it exerts its autonomy. These transformations in turn modify the agent's room for maneuver. They constrain it, while at the same time giving the robot or organism new opportunities for action ("affordances," in technical parlance). In other words, every autonomous entity inevitably creates a niche for itself in which its autonomy can be securely anchored. This is why the autonomy of McFarland's lawn mower is more metaphorical than real: the lawn mower not only does not create a niche that allows it to exist, it does not even maintain a niche. It receives its autonomy ready-made from the outside. Its dependence on the external world, and especially on whoever is responsible for looking after it, is complete. The lawn mower transforms its environment only in a very limited way. Cutting the grass and recharging its batteries as needed—and never doing anything else—constitute the whole of its reason for being.

A social robot, if it is to be able to fit in with its environment, must have a certain flexibility in its relations with human beings. The *substitute* must be able to create its own niche. In order to do this, it needs neither energy autonomy nor motivational autonomy. It is also why it cannot be compelled to perform one and only one function or to carry out one and only one program of action. By "flexibility" we mean that a social robot, unlike a door closer or a turnspit, is capable of interrupting or abandoning a particular activity and adjusting or reconfiguring its relations with its human partners accordingly. This ability has two aspects. First, the robot must be able to react appropriately in the event that a particular activity is interrupted or canceled by its partner. Second, the robot must itself be able to stop or abandon an activity in which it is engaged in response to external events of a certain sort. Yet the disjunction, or discontinuity, caused by interfering with the course of an activity in either case cannot be in any way arbitrary or unmotivated; in order for reciprocal coordination to be possible, the robot's response must correspond to some recognizable regularity.[29] For correspondence to obtain in this case, the disjunction need not express or refer to a preexisting behavioral regularity; but it must be such that it will be possible in retrospect, considering the interaction in its entirety, to detect a regularity of which the disjunction can henceforth be regarded as the "expression."[30]

Autonomy in this sense is in one way very much an individual characteristic: some agents are autonomous; others are not. But social autonomy nevertheless remains a relational property that exists only in relation to a setting that serves as a frame of reference. An agent may be said to have social autonomy only if it belongs to a population of agents that constitute the environment within which the agent acts. This environment, just like the lawn mower's environment, must be appropriate, which is to say that it must be organized in such a way as to allow the robot to exercise its autonomy. But this environment, unlike that of the lawn mower, is—to use the vocabulary of game theory—strategic rather than parametric. It is an environment that is capable of being rapidly and significantly transformed in response to the robot's action, for the environment is chiefly composed of the social partners with whom it interacts. These sudden environmental changes, though they may prevent the robot from attaining a desired end, also offer it new possibilities of action.

A robot that is permanently attached to an external energy source, no less than an immobilized patient hooked up to an artificial lung, is deprived of what McFarland calls energy autonomy. A human slave and a semiautonomous robot—that is, a machine that is remotely controlled by a human operator—are both deprived of what he calls motivational autonomy. All of them are nonetheless capable of exhibiting social autonomy if they can continuously coordinate their behavior with that of their partners in response to changing conditions of interaction.[31] Social coordination is not merely a matter of being able to accomplish a particular objective—going to the movies or unloading a truck. It is a matter also of being able to repeatedly interact in different ways, for a wide variety of reasons and with a wide variety of aims. Social coordination therefore requires not only that agents be capable of coordinating their actions as a function of a specific task, but also that they be able to coordinate their actions over an open-ended set of distinct tasks. If they satisfy this dual condition of social coordination and re-coordination—as even slaves evidently do—then they are socially autonomous. Although social autonomy pertains to a particular dimension of action, it is not itself a normative concept, nor does it entail any metaphysical claim concerning the agent's liberty.[32]

Robots as Scientific Instruments

"It would seem that at least three things characteristically human are out of reach of contemporary automata," John Cohen observed more than fifty years ago. "In the first place, they are incapable of laughter (or tears); secondly, they do not blush; thirdly, they do not commit suicide."[33] Not only does this remain true today, it is all the more timely to recall as emotions have in the meantime come to be a main focus of research in cognitive science, after a long period of neglect dominated by the classical paradigm of the brain as a symbol-processing machine. Two recent approaches in the theory of mind, known as enaction and embodied mind, assign a central place in human knowledge to the body and to emotions, which previously

were either considered not to be cognitive or else regarded as obstacles to rational thought.

One of the difficulties in creating artificial agents that resemble
human beings arises from the fact that while we have a good understanding of the physiological mechanisms involved in blushing, for
example, we do not have a very clear idea under what circumstances
and for what reasons people blush. At most, we know how to answer
such questions only in an anecdotal fashion. The four distinctively
human behaviors that Cohen mentions—laughing, crying, blushing,
and committing suicide—are also preeminently social behaviors.
They play a fundamental role in all human phenomena of association
and disassociation. We often laugh together, and sometimes we laugh
at others; we blush in the presence of others; we cry and even commit
suicide in response to what others have done to us or of what has
happened to them. But if we are all capable of these behaviors, as
Cohen points out, we do not really understand what they mean, what
social function they serve, or what causes them to be exhibited. The
plain fact of the matter is that many aspects of human sociability remain mysterious to us. What holds a human group together? What
inclines a person to persist in a plan of action or to abandon it?
What encourages learning or stands in the way of it? What enables us
to get along with others and feel comfortable with them?

Many roboticists regard the artificial social agents they create as
scientific instruments—research tools, means of advancing knowledge about the nature of sociability. They hardly suppose that they
are presently in a position to answer the preceding questions. But
they do think that the development of robots of a certain type will
allow significant progress to be made in this direction. Their reasons
for optimism are essentially two. In the first place, introducing a robot
in the company of human beings—in a retirement home, a school, a
hospital, or a supermarket—is an inherently social phenomenon that
can be treated as a scientific experiment. Studying a whole range of

reactions associated with acceptance and rejection—anger, annoy-
ance, amusement, affection—and their short- and long-term effects
cannot help but teach us many things about various aspects of human
sociability, even if it is not always clear how to interpret the information that is collected.

The three-dimensional physical presence of a robot radically
distinguishes it from merely virtual agents, which only appear on
projection screens in a visual environment already saturated with
them. Virtual agents can play their role only if the spectator is willing
to let them—if what he sees on the screen interests him. A major
problem for researchers who rely on virtual agents is precisely this,
how to capture the attention of an audience that often is indifferent
to begin with or distracted, if not actually preoccupied, by any number
of visual appeals and auditory requests transmitted on competing
screens. A robot, by contrast, imposes itself. It intrudes on the very
physical space we occupy. Its presence is all the more difficult to ignore, or to pretend to ignore, as we are faced with a complex social
agent that to one degree or another is capable of taking us as an object of its attention. Whereas our interactions with virtual artificial
agents can usually be analyzed without difficulty in terms of the relationship of an audience to one of the performing arts or of the public
to the media, we interact with robots in an altogether different fashion.
There is no uncanny valley for virtual agents, for they remain confined to a screen; they do not thrust themselves upon us, in the literal
sense of entering into the three-dimensional physical world that we
inhabit.[34] The introduction of a social robot is an event of a very particular type, one that more closely resembles the appearance of an
anthropologist among an unknown people than the introduction of
a new communications medium.

A second reason for optimism relates to the fact that creating an
autonomous social robot requires it to be capable of human behaviors that we do not yet fully understand ourselves. We do know very

well, of course, how to create virtual agents that laugh or cry, that express shame or indeed any other emotion, appropriately and in a convincing fashion; or that make an audience laugh, or cry, or express intense feeling in another way. These agents are usually fictional characters in an entertainment of some kind. They play an important role in our lives—disturbingly important in the view of those who fear that we may lose sight of the distinction between reality and imagination. But these agents are also disembodied. Implementing the same behaviors in embodied artificial agents that can take part in real social interactions presents challenges that are as much ethical as technological. It is not hard to draw up a plausible and more or less exhaustive list of the circumstances in which people cry, laugh, blush, or commit suicide. But to be able to write, on the basis of such a list, a program that will permit a robot to act in a socially convincing and morally acceptable fashion is by no means easy to do.

· This is why each prototype, and each trial, constitutes an experiment, a means of learning a little more about behaviors that, if the roboticists are not mistaken, we will not truly begin to understand until we know how to give them practical expression in an artificial agent. Note that two things are implied here: first, that the ability to give expression to these behaviors in robotic agents is what will enable us to truly understand them; second, that it is only to the extent we truly understand these behaviors that we will be capable of giving expression to them in artificial autonomous agents. It is precisely in this dual sense that social robots need to be considered as scientific instruments. The more successful we are in creating robots that exhibit typically human behaviors, the better we will understand these behaviors and the roles they play in human sociability. And the better we understand these behaviors and the roles they play in human sociability, the more successful we will be in implementing them in artificial agents. This is a virtuous, not a vicious, circle.

The scientific experiment being conducted by social robotics, particularly the branch of it that aims to fabricate substitutes, is not without consequence for a longstanding debate in philosophy of mind and cognitive science about the nature of the human mind. This debate opposes what is generally called classical cognitivism with one or another version of the theory of embodied mind,[35] from "simple embodiment" at one extreme to "radical embodiment" at the other.[36] In the classical view, the computer recommends itself as the proper metaphor of the mind, because it allows us to see the mind as a truly universal computational capacity that can be realized in radically different physical supports, whether biological neurons, integrated circuits, or the artificial neurons of neural network models. This approach conceives of the mind as something disembodied, as a "naked mind"[37] reduced to a pure intellectual object—a computer program, for example, that can be written and implemented and then run on different types of machines, perhaps even copied and saved to a disk where it can live on in another form after our death.[38] In contrast, the new view holds that the mind cannot be confined within the brain; it extends beyond it, becoming incarnate not only in our bodies, but also in our tools and in our technologies.[39]

The remarkable thing about social robotics, as a scientific experiment, is that it does not limit itself to taking a position in the debate, hoping to tip the scales in favor of one side rather than the other. It alters the very terms of the debate as well. Both its successes and its failures illustrate the socially distributed character of cognitive behavior. Paradoxically, or perhaps not, it suggests that the social distribution of mind is what also allows thinking to be a solitary and solipsistic activity,[40] just as it allows the behavior of computers to be considered cognitive.

The social distribution of cognition is a kind of doing or making. It transforms the world—the social world, to begin with, but not only the social world, for it transforms the physical environment as well. By changing the world we live in, social robotics will reveal some of its most basic features, while also accentuating them and making them more obvious to us. So far as social robots are a means of learning about ourselves, again this is because introducing artificial agents, some of them autonomous, will recast the nature of our relationship to our fellow human beings, allowing us to know each other better. Artificial social actors will change the world in ways that we cannot predict and do not yet understand, since the transformations of which they will partly be the cause constitute the very process by which we will come to have a better understanding of our world.

In fact, there is nothing paradoxical about any of this. Gaston Bachelard showed long ago that modern science discovers the world by transforming it. Science does not simply "save" phenomena. It creates them. It enlists them, trains them, disciplines them, makes them amenable to our will. As Gilbert Simondon put it, the artificial is induced nature.

CHAPTER 2

Animals, Machines, Cyborgs, and the Taxi

When two, or more men, know of one and the same fact, they are said to be CONSCIOUS of it one to another; which is as much as to know it together.

THOMAS HOBBES

Artificial ethology encompasses a theoretical perspective and an empirical body of research in which robots are used to understand, analyze, and reproduce experimentally certain aspects of animal behavior.[1] It is not interested solely, or even mainly, in creating biomimetic robots ("animats," as they are called)—robotic butterflies, for example, or dogs or snakes. Its chief aim is to fabricate creatures that, though they may bear only a distant resemblance to a particular animal, are capable of implementing or exemplifying certain behaviors that are found in animals. Take the problem of constructing a robotic lobster that can turn toward the source of an odor diffused in turbulent water currents, as real lobsters do when they search for mussels or clams. It is remarkable that lobsters do not all die from hunger. Because of the turbulence, there is no simple and direct correspondence, no linear relation, between the density of the odor and the distance from the source; what is more, the odor reaches the lobster only in short intermittent gusts, rather than in a continuous fashion, and from directions that may sharply vary even when the source is stationary. A similar problem is encountered in trying to build a robot capable of recognizing and orienting itself toward the song of a male cricket of a particular species in an environment in which are present several males of closely related species whose songs are only

58

very slightly different. How does a female cricket manage to turn toward a male of the right species when she does not see him, only hears him—and this in an environment in which his song is mingled with that of other males? What cognitive resources must she be able to bring to bear to be able to recognize the song of a member of her species among so many similar ones?

Artificial ethology hopes to be able to answer classic questions of this sort by means of robots. These are not social robots, still less substitutes; rather, they are scientific instruments, research tools whose sole useful purpose is to assist the advance of knowledge and the development of new technologies.[2] The researcher's aim is to construct an artificial agent capable of solving a problem faced by its animal counterpart. Making a robot that is able to satisfactorily reproduce the behavior being studied requires figuring out what resources the animal can exploit, given the constraints imposed by its morphology and the nature of its environment, which in turn gives us a better understanding of how the animal manages to perform a given task. Robotic modeling makes it possible to carry out experiments in both cognitive ethology and philosophy of mind with a view not only to testing traditional assumptions, but also to devising new hypotheses.[3]

Looking to robots for answers to these questions has a number of advantages compared with more speculative classical approaches. The most important one is that it now becomes possible to decide if a given explanation for a certain type of animal behavior is plausible, or in fact likely to be true, by implementing it in a robotic system. To be sure, reproducing an animal's behavior by means of a model cannot by itself *prove* that a hypothesis is correct, any more than any other scientific experiment can, considered in isolation. It is also true that the robotic approach has its limits, due to the specific type of modeling involved.[4] Even when a robot succeeds in reproducing an animal's behavior—as in the preceding examples, by turning toward the source of an odor or a male of the right species—one cannot be sure that the animal performs this function in the same manner, exploiting the same cognitive resources. Nevertheless, one can say whether or not the hypothesis is plausible, and it will be corroborated all the more if the robot's behavior manifests the same type of errors and the same limitations as the animal's behavior. As Frank Grasso notes, robotic modeling often proves to be instructive even when it fails, in as much as it can show that a commonly accepted explanation of some aspect of an animal's behavior is, in fact, false. This happens when robotic modeling establishes that the animal cannot reach its intended goal, once the particular conditions it faces and the resources that are available to it have been properly considered.[5]

The close attention paid to morphological and environmental constraints is fundamental and accounts in large part for the appeal of robotic modeling compared with computer simulation, whose scenarios concern virtual agents acting in a virtual universe. The use of robots forces experimenters to reckon with an actual environment, within which an animal's behavior unfolds in an entirely different manner. The virtual environment assumed by computer simulation constrains the behavior of virtual automata by means of parameters. The modeler isolates those aspects of the animal and the environment that he considers pertinent, incorporates the hypotheses that he wishes to test, and runs the model. Any aspect of the environment not contemplated by the model simply does not exist in the universe in which a logical automaton tries to reproduce the animal behavior in question. A robot, by contrast, as a three-dimensional object in physical space, moves around in a much noisier and more chaotic environment, which is also more similar to the one in which the animal acts. To be sure, very often it is an artificial experimental environment, a controlled setting that is rather different from an actual animal environment. But the robot's physical structure and its real

presence require that the modeling of behaviors and environments include a great many constraints that are liable to be omitted
in a computer simulation. Where a decision has been made not to
take something into account, it is generally carefully considered and
biologically informed, though it may also recommend itself for reasons of simplicity or because it is impossible, or too complicated, to
represent this or that aspect of the environment mathematically.
Because the robot operates in a real, rather than a virtual, space, very
often one simply cannot avoid dealing with comparatively intractable elements of a problem, with the result that it is much less likely
that an incorrect modeling of fundamental dimensions of the environment will go unnoticed.[6] Robotic simulation also makes it possible to appreciate the extent to which the environment is a source
not only of constraints, but also of opportunities that an animal can
seize upon to its own benefit. Some of these affordances are the result of fortuitous encounters that are very difficult to predict and to
characterize analytically in advance, but that are often revealed when
the experimenter looks at a real physical object that imitates the behavior of an animal in a sufficiently similar environment. Robotic
modeling facilitates the discovery of these unforeseen positive interactions, whereas simulation on a computer tends to conceal them.[7]

  If robotic modeling can show that a certain hypothesis advanced
to explain a particular behavior will not give the anticipated result, it
may also sometimes show, to the contrary, that what was considered
by hypothesis to be impossible in principle may in fact occur. This is
especially true with regard to hypotheses concerning cognitive resources that are supposed to be necessary to exhibit this or that behavior. The difficulty of testing such hypotheses with the aid of real
animals comes from the fact that one cannot very well assert that
they do not have the cognitive abilities in question. If an animal manifests this or that behavior, it is, by definition, because it possesses the
cognitive abilities needed to implement it. Yet this tautology does

not tell us exactly which cognitive abilities are actually being exercised. It does not tell us if the ones we postulate are in fact the ones responsible for the behavior. Robotic modeling may allow us to break the hold of this circular reasoning by making an artificial agent capable of displaying the behavior we are interested in, but without having to appeal to the same hypotheses regarding the cognitive abilities needed to produce it.

A closely related field of research in which problems of this kind are also encountered, and which we take up in greater detail in Chapter 4, is concerned with what it means to have a "theory of mind." An experimental protocol very commonly followed in cognitive psychology uses the attribution of false beliefs to others as a basis for determining the age at which children acquire such a theory. The ability to attribute false beliefs is taken as a sign that children are able to regard others as agents having complex mental states that structure the way they perceive the world. Even if the experimental results obtained are fairly robust on the whole, their interpretation remains problematic. In the first place, it is extremely difficult to determine if the conceptual resources deemed necessary to attribute false beliefs are really and truly necessary. Second, it is no less difficult to determine whether the resources that are in fact exploited by children (and adults) are indeed the ones that are assumed to be operative. Third, the question arises whether these conceptual resources, and the ability to ascribe complex mental states to others that they make possible, were not already present in the child long before their presence was revealed by experiment. In that case, the appearance of an ability to attribute false beliefs at a particular age would have an altogether different significance than it is generally supposed to have.[8]

The problem is that the number of possible replies to such questions seems to be limited only by the researcher's imagination. Once again, making use of robots wherever possible will very often show a way out from the quandary. If a robot proves to be capable of manifesting the ability in question, we can sometimes decide whether it in fact has the cognitive resources commonly judged to be a necessary condition of behaving in a certain way, for we know—since we have constructed it—whether the robot is able to represent to itself the particular aspect of the world that the behavior in question is thought to require. Even if the result of a robotic experiment does not suffice to show that the child or animal is able to produce a certain behavior by drawing upon the postulated resources, it may be enough to demonstrate that what had been held to be a logical impossibility is not in fact one, and that the behavior can be explained more economically, with the aid of a hypothesis that demands less in the way of cognitive or computational investment.[9]

An Experimental Philosophy of Animal Mind

Artificial ethology therefore constitutes an experimental philosophy of animal mind. At the methodological level, reliance on robots to explain animal behavior helps in at least two ways to advance research in both cognitive ethology and philosophy of mind. On the one hand, it acts as a filter, making it possible to do a better job of evaluating the soundness of hypotheses whose explanatory power has long been disputed. On the other hand, it is a source of new hypotheses that can immediately be tested. Fabricating robots that can function in real, rather than simulated, environments requires first of all solving a whole set of technical problems connected with the constraints imposed by what is assumed to be possible in the real world. The solutions proposed to these problems constitute hypotheses in their turn, which, once tested, will give us a better sense of what in fact is possible in the real world. Constructing one or more robots is a way of testing a hypothesis, of refining it and making it more precise. This threefold purpose—identifying difficulties, forming a clearer idea of the problem to be solved, and putting forward more promising hypotheses for research—has characterized the use of experimental techniques in the sciences since Galileo.

Reliance on robots to simulate and understand animal behavior is compatible with the embodied mind approach in philosophy of mind and cognitive science[10] (particularly the tendency known as "radical embodiment"[11]), as well as the "enactive approach,"[12] with which some researchers in artificial ethology have openly aligned themselves.[13] Even if, as is often the case, a particular ability is modeled in isolation from the rest of the animal's behavior, the results obtained so far suggest that the means implemented by different animal behaviors exceed what an animal's internal cognitive resources are alone capable of supplying. Explaining the cognitive behavior of an animal requires taking into account the physical characteristics of the animal's body and its environment. The animal mind is not purely mental. It is not a "naked mind" that can be reduced to a set of algorithms or to a particular computational architecture; it is situated, which is to say inseparable from its surroundings.[14] Animal cognitive abilities can therefore be seen to lend support to the thesis of embodied mind: they cannot be measured solely in terms of the cerebral—neuronal and computational—resources available to a particular animal; they result instead from the interaction of these internal resources with the resources supplied and the constraints, imposed as much by the animal's body as by the ecological niche that the animal occupies.

Artificial ethology also contributes to what might be called comparative cognitive science and philosophy of mind, and this in at least two ways. First, by illustrating the diversity of embodied animal intelligence through robotic modeling, it shows that the intelligence of an animal agent, or of an artifact, is closely tied to its physiology and to a particular environment. Second, by actually making robots, it produces artificial cognitive systems capable of reproducing certain aspects of animal intelligences. The dependence of animal mind on the organism's body and its environment, which artificial ethology exploits and tries to reproduce, together with the ability to reproduce some, though not all, of the cognitive capacities of animals, suggest an argument that we develop later and that will be a central part of what we are trying to establish in this book. This argument remains implicit for the most part in "the embodied mind approach," and only seldom is it asserted explicitly, namely, that there exist several forms of "mind," which is to say several types of cognitive systems. The cognitive domain, we maintain, is much less homogeneous than is generally assumed. There are many ways of being cognitive; and the robots devised by artificial ethology themselves are cognitive systems, but of a different type than the systems of the animals whose cognitive abilities they model.[15]

It may well be, then, that cognitive systems of distinct types resemble one another more in the way that distant relatives in a phylogenetic tree do than individual members of the same species. Researchers who identify life with cognition, such as Humberto Maturana and Francisco Varela,[16] Evan Thompson,[17] and Michel Bitbol and Pier Luigi Luisi,[18] all of them exponents of the embodied mind approach, tacitly accept this hypothesis for the most part. Sometimes they openly acknowledge it, as Maturana and Varela do in their book *The Tree of Knowledge*, which seeks to retrace the phylogenesis of various forms of natural cognition.[19] If being a living system and being a cognitive system are inseparable aspects of a single reality, if every living system is necessarily a cognitive system, why should there not be as much diversity among cognitive systems as there is between an amoeba, a conifer, and a horse? The question is all the more pertinent since plainly there also exist cognitive systems that are not living systems.

Artificial ethology leads us to take a fresh look at the nature of both animal and human minds, as well as their relations with artificial cognitive systems that are capable to one degree or another of imitating the cognitive capacities of these different types of minds, and capable to one degree or another of doing what human and animal minds are incapable of doing. First, however, the hypothesis of the diversity of cognitive systems, in combination with experimental results from artificial ethology that show the animal mind to be irremediably local, embodied, and situated, invites us to revisit a much older philosophical thesis, itself part of the first modern formulation of the philosophy of mind: the Cartesian doctrine that animals are machines.

The Animal-Machine

Closely connected with Cartesian dualism, the animal-machine thesis has generally been misunderstood. Rejected in Descartes's own time as implausible, today it is often considered to be both false and self-evidently true. The argument Descartes makes is, in any case, more subtle than is commonly supposed. It throws valuable light on current debates in both philosophy and cognitive science, for the ritual and indeed almost automatic dismissal of Cartesian dualism frequently goes hand in hand with a spontaneous uncritical adoption of the very conceptual structure that Descartes imposed on the problem of understanding the human mind in the *Discourse on Method*. As a consequence, and despite its avowed monism, contemporary thinking remains to a large extent Cartesian.

It is important to recall that Cartesian dualism was intended, among other things, as a refutation of vitalism. In rejecting the idea that there exists a "living matter," different in kind from physical matter and subject to laws other than the ones that govern the whole of the physical universe, it aimed at creating for the physical sciences a unified domain of research.[20] The animal-machine thesis is an immediate consequence of this methodological program. There is a consensus today that living beings are complex machines endowed with a particular form of organization, without there being any mysterious vital principle that distinguishes organic molecules from inanimate matter. In this regard, the truth of the Cartesian thesis is taken to go without saying.

According to Descartes, only the human mind—not the body— stands apart from the empire over which physical science otherwise reigns absolutely. This distinction is now generally rejected. There is no reason, it is said, to exclude human beings, and more particularly the human mind, from the unified domain over which the sciences exert their authority. Modern technologies show that it is possible to construct material cognitive systems. One of these in particular—the computer, the universal Turing machine—has for at least half a century now been the preeminent model for understanding the human mind.[21] Furthermore, ethology has revealed the richness and complexity of cognitive behaviors in animals. In this regard, then, the thesis of animals as machines is considered to be false for two reasons: on the one hand, because it rests on a discredited dualism; on the other, because we take for granted that it denies the existence of cognitive abilities in animals. It is believed, in other words, that Descartes's refusal to recognize that animals have a soul and his willingness to grant mental status only to human beings mean that he considered animals to be devoid of cognitive ability altogether.

But the animal-machine thesis does not entail either that animals have no cognitive abilities or that animal behaviors that seem to be cognitive are not really cognitive. To the contrary, Descartes is perfectly aware that animals manifest many cognitive behaviors and that they sometimes are better than we are at doing what we do, but he is attentive also to the fact that the cognitive abilities of animals are different from ours. In the *Discourse on Method* (1637), he makes a point of emphasizing that animals have cognitive abilities, even though they may differ markedly from the ones he seeks to describe as being peculiar to human beings:

> It is also something very remarkable that, although there are many animals that show more skill [*industrie*] than us in some of their actions, one still notices that the same animals do not show any skill at all in many others. Thus whatever they do better than us does not prove that they have a mind [*esprit*] because, on this assumption, they would have more intelligence [*esprit*] than any of us and would be better at everything. It proves, rather, that they have no intelligence at all, and that it is nature which acts in them in accordance with the disposition of their organs, just as we see that a clock, which is made only of wheels and springs, can count the hours and measure time more accurately than we can with all our efforts.[22]

In refusing to grant mindfulness to animals, Descartes does not deny them cognitive ability altogether. He does suppose, however, that the animal's cognitive behavior is different in kind than human cognitive behavior, which alone depends on what he calls mind. Descartes holds that the cognitive abilities of animals are susceptible to the same type of explanation as the cognitive behavior of machines. As the philosopher Thierry Gontier points out, "If animals are not reasonable beings, they nonetheless remain (just like machines) rational beings, moved by a reason that directs them toward the end assigned to them."[23] This interpretation implies the existence of purely material cognitive systems, which is to say ones without soul or mind, and suggests that Descartes would have been filled with admiration, rather than shocked or in any way dismayed, to discover the many material cognitive systems found in the world today. What Descartes refuses to recognize in animals is exactly what we refuse to recognize in the cognitive systems that we fabricate ourselves, namely, thought, language, and reflexive consciousness. Nevertheless, he does not imagine that animals are wholly lacking in cognitive ability.

Moreover, according to Descartes, animals are machines of a very particular type, quite different from the ones that we build, for they have been created by a master artisan whose skill is far superior to any that human beings can summon. Animals must therefore be considered *methodologically* as machines, so to speak, in the sense that nothing in their behavior suggests that they are amenable to a type of explanation different from the one that we apply in creating and understanding machines ourselves. Animals are machines in the sense that the machines that we make permit us to imagine what a maker infinitely more skillful than ourselves could create on the basis of the same principles that guide us when we build ours.

In the Cartesian view, there is no reason to think that animals are not machines, which is to say no reason to think that they have a soul, because their cognitive abilities, real though they undoubtedly are, are very different from ours. Whereas animals, just like machines, perform certain purely cognitive tasks better than we do, there are nonetheless other tasks that, again, just like machines, they are wholly incapable of performing. When human beings, machines, and animals are faced with the same problem—measuring time, for example—they do not commit the same types of errors. Cartesian dualism distinguishes between mind, which is peculiar to human beings, and the cognitive abilities of animals and machines. Yet it does not entirely identify animals and machines with one another; animals are indeed machines, but they are cognitively much more efficient than any machine we know how to make. What the theory of animals as machines asserts is that there are different types of cognitive systems, and that the distribution of cognitive errors and successes among them makes it possible to distinguish these different types of cognitive systems from one another.

This interpretation of the animal-machine thesis suggests another reading of Cartesian dualism, apart from the fundamental distinction it makes between two types of cognitive systems. Its appeal consists in the fact that it recognizes the existence of a plurality of cognitive systems. Contemporary monism refuses to entertain the idea of pluralism in this connection because, on the one hand, it fails to see that the separation instituted by the Cartesian mind-body dichotomy occurs within the cognitive domain and, on the other, because it thinks that the rejection of dualism excludes any possible discontinuity within this domain. But the repudiation of cognitive pluralism places philosophy of mind in the untenable position of reaffirming a difference that it denies and presupposing a distinction between the mind and artificial systems that it rejects. Artificial ethology, social robotics, and certain theories of embodied mind have now put cognitive pluralism back in play. They argue for replacing the abstract alternative between dualism and monism, which is often conceived as a choice between spiritualism and science, with the theoretical analysis and empirical investigation of the diversity of cognitive systems.

On Cognitive Pluralism, or the Diversity of Minds

According to Descartes, what fundamentally sets human beings apart from animals and machines is the understanding and use of language. It is this ability that constitutes the absolute criterion for telling the difference between the two types of cognitive systems that his dualism is meant to keep separate. Notwithstanding the repeated and ritual condemnations to which Cartesian dualism is subject, however, not only does this separation still have a central place in cognitive science and philosophy of mind today, but it continues to play the same role in them: distinguishing between "true" intelligence, which characterizes human beings, and the intelligence that characterizes the machines we make, whose behavior is explained, as Descartes put it, by the disposition of their organs. To be sure, these disciplines take it for granted that typically human cognitive abilities will one day be explained in the same way that those of machines are, but for the time being the difference between them persists. It remains at once inexplicable, indispensable, and yet doomed to disappear, to be eradicated by the advance of knowledge.

Contemporary philosophy of mind has retained from Descartes much more than it is prepared to admit. In particular, it has kept the conceptual structure that he imposed on the problem of mind, notably the essential role assigned to the understanding of language. As a consequence, it finds itself encumbered by a distinction that it does not know how to make sense of. The criteria employed vary, but in every case they are used to distinguish human intelligence from other simpler and different (and so less "true") forms of cognition. Sometimes it is a matter of distinguishing between general intelligence and modular systems,[24] sometimes between consciousness in a strong sense and other weaker forms,[25] or else between intrinsic intentionality and derivative intentionality,[26] or between intrinsic content and derivative content.[27] In spite of the diversity of criteria, the underlying idea remains the same: the human mind, which is uniquely capable of understanding language and of giving meaning to things, differs from all other cognitive systems we are acquainted with, whether natural or artificial. It possesses certain characteristics that we do not know how to reproduce, at least not yet, and that moreover are not met with in the natural world. Nevertheless, there is a firm and widely shared conviction that we will one day succeed in bridging the gap between human and artificial intelligence. What is holding us back, it is believed, are the momentary limitations of our understanding, not any real and permanent separation between distinct types of cognitive systems.

Cognitive scientists find themselves today in a paradoxical situation, namely, that they recognize in practice a distinction that they refuse to recognize in principle. On the one hand, both philosophy of mind and cognitive science reject Cartesian dualism, rhetorically at least; on the other, they affirm the importance of the distinction on which it is based. In other words, they take to be primary a difference that they claim ought not to exist or, more precisely, that should not have the weight that is accorded to it—not only because the difference between human and machine intelligence is destined one day to disappear (the day when machines become conscious!), but also because, independently of the progress of our knowledge and of our technologies, it is supposed to be a difference of degree rather than of kind. Nevertheless, as we will soon see, a number of prominent philosophers of mind use this very distinction between two kinds of intelligence, which as monists they must officially and explicitly reject, as a criterion for determining what really counts as mind and cognition. Recognizing the heterogeneity of cognitive faculties requires that we not deny the distinction between the human mind and animal or machine intelligence *and* that we not transform it into an ontological gulf. To do both these things it is enough to exchange a dichotomy for a pluralism—one that accepts, among other things, the legitimacy of this distinction insofar as it marks off some of our cognitive abilities as peculiar to a particular type of cognitive system. Social robotics, and especially research on the class of technological objects that we call substitutes, can help us understand and explain at least a part of the difference between human and other forms of intelligence, which philosophy of mind and cognitive science refuse to acknowledge even as they hasten to exploit it. Social robotics may be thought of, in something like the manner of artificial ethology, as a kind of artificial anthropology whose aim is to recreate and produce certain typically human cognitive abilities, distinct from the ones that the computer seeks to extend, deepen, and accelerate.

Robotic modeling does more than reproduce and simulate. It implements as well. To do this, the active participation of one or more human partners is required. Participation in this case is not limited to understanding or interpreting what the machine does or says. It constitutes a moment within a process that is distributed among several agents, of which the human partner is neither the sole author nor the owner. Here we find another form of plurality, not a plurality of cognitive systems, but a plurality of actors taking part in a cognitive process. This form of participation, we believe, underlies the functioning of what is usually called mind.

The Extended Mind and Cyborgs

Almost twenty years ago now, the philosophers Andy Clark and David Chalmers sketched a theory of what they called the extended mind in a famous article of the same name.[28] The expression "extended mind" is to be understood both in a metaphorical sense and more strictly. In the metaphorical sense, the extended mind is an augmented mind, improved through the application of technologies that, from the invention of writing to the advent of the Internet, have multiplied and increased our cognitive powers. But the extended mind is also, and especially, extended in a literal sense, which is to say that it is extended in space. If mind is, properly speaking, something material—as the theory of property dualism would have it[29]—it is an object situated in a definite place in physical space. It is a material object that has the astonishing property of possessing mental states, which correspond to—or, to use the customary technical term, supervene on—physical processes arising from the concatenation of various parts of this object. This object, the mind, has a spatial extension that is generally held to coincide with the area occupied by the human brain. Clark and Chalmers argue, however, that when cognitive processes include or mobilize objects that are external to the brain itself, there can be no justification for limiting the mind to the brain.

The thesis that the mind extends beyond the brain and into more or less adjacent physical space is at once an empirical claim, about the role of intellectual techniques and cognitive technologies in explaining and expanding human cognitive abilities, and a metaphysical claim. In its metaphysical aspect it asserts that, if one accepts that the mind is an extended object in physical space, then it is incoherent to say that it is extended, in the sense that its power is augmented or enhanced by reliance on physical objects that are external to it, unless one also recognizes that by virtue of just this the mind becomes more extended in the strict sense, which is to say in physical space. The extended mind thesis may therefore be seen as the consequence both of adopting a materialist perspective on the mind and taking a certain view of the empirical role of various technologies in the development and functioning of the human mind.[30]

By way of illustration, Clark and Chalmers ask us to consider the predicament of a man named Otto, who is suffering from Alzheimer's disease, and his friend Inga. Both of them live in New York. Otto wants to see an exhibition at the Museum of Modern Art, but he no longer remembers the address of the museum and so consults his trusty address book. Inga, who wants to see the same exhibition, simply relies on her biological memory. In each case, the mental process is the same, remembering the museum's address, but it is carried out with the aid of different physical supports. In Inga's case, the mind's physical support consists of her brain alone, the seat of biological memory; in Otto's case, the mental act of remembering the address of the Museum of Modern Art depends both on his brain and his address book, which supplements his failing (biological) memory. Clark and Chalmers maintain that in certain cases, where the material supports of cognition are situated in part outside the individual's brain, these external supports are properly said to be part of one's mind, which therefore extends beyond the brain. When, as in this case, there exists a functional equivalence between an external process and an internal process, there can be no reason to refuse to recognize this external process as constituting an integral part of the agent's mind. According to Clark, this capacity of extending— stretching, leaking, seeping—beyond the limits of the skull and the skin is characteristic of the human mind. As a result, he says, we are all natural cyborgs.[31]

What Clark fails to notice is that the "natural" cyborgs he imagines human beings to be are essentially "metaphysical," disembodied cyborgs. The mind's ability to extend itself beyond our bodies, as he describes it and defines it, requires no implant, no mechanical augmentation that ruptures the physical integrity of the individual person. A good book, or even an address book, will do perfectly well. Human beings, in other words, are *intellectual cyborgs*, for the unity of the physical processes that are supposed to constitute the extended mind does not exist. At the physical level, we are dealing with utterly separate, independent processes. There is nothing wrong, of course, with supposing that a certain continuity, or unity, obtains among the neuronal processes that occur in my brain when I remember an address. But it is difficult to imagine what connection there might be between the physical, which is to say molecular, processes underlying the visible features that constitute an address written down on paper and the processes that occur in my brain when I read this address. It is quite certain that something goes on in my brain when I read an address, but what goes on in my brain at the physical level has nothing to do with what constitutes the address as a physical object; it is only what goes on in my brain as a mental state or process that has something to do with the address as a conventional sign. There is no physical unity between the physical processes that, according to Clark and Chalmers, form the extended mind; there is simply a mental relation, a relation of signification between the thought of the Museum of Modern Art and the address that Otto finds written down in his address book. Strictly speaking, in the example they give, mind is not so much extended as *distributed*. A given mental process—remembering an address—is realized by a set of physical dynamics that are very different from one another, among which there is no unity on the physical, spatial level.

Clark and Chalmers remain captives of the dualist alternative as Descartes defined it, for they believe that incorporating, embodying the mind means assigning it a corporeal seat, a definite place, and confining it in a physical dimension conceived as purely spatial extension. This makes it impossible to account for the dynamic and processual aspects of the mind, which they are obliged to reduce to a mere functional equivalence—an equivalence that erases fundamental mental and physical differences between Inga, who remembers, and Otto, who reads, which is to say learns and discovers, the address of the Museum of Modern Art. Their approach leads back, ultimately, to an abstract and disembodied conception of the mind.

The extended mind thesis has given rise to important debates in philosophy of mind.[32] They bear essentially on the question of whether it is legitimate, and, if so, under what circumstances, to hold that cognitive processes that occur partly outside a person's body may nonetheless be said to belong to his mind. The issue here has to do with the status of external aids to understanding that make us more intelligent and intellectually more capable. If a tool both augments and determines—if it informs, in the Aristotelian sense—the ability of a craftsman, we do not generally consider that the tool belongs to the body of the person who uses it, except when we are dealing with a true cyborg. Why should the situation be any different here? Why should these technologies, which are not a part of our body, be a part of our mind? How do they extend it materially?

It must be said, to begin with, that philosophers of mind are indebted to Clark for having reminded them of the central place occupied by technology in the development and the functioning of human cognitive abilities. Even cognitive science tends to forget their importance—a most curious thing, since computers and artificial neural networks and the like are precisely what made this discipline possible in the first place and enabled it to thrive! And yet the crucial role he assigns to artificial aids to understanding turns out on closer examination to be rather ambiguous.

The extended mind thesis postulates a very particular relationship between the mind and the various material technologies of cognition to the extent that these technologies are understood as extensions of the mind in a strict sense or, to put it another way, to the extent that the mind is thought of as a natural cyborg. The extended mind thesis affirms that these technologies and the objects that materialize them, from the earliest forms of writing to the most powerful computers, augment, enhance, enlarge, and actually extend the human mind. This amounts to saying that the human mind absorbs these technologies, in the sense that all the cognitive systems with which it interacts become part of the mind with which they interact. They are assimilated by the mind. Accordingly, the extended mind thesis grants no autonomy, no independence to these technologies; they are genuinely cognitive only insofar as they are appendages of the human mind, which, however extended it may be, continues to have its seat in individual brains. For this reason, the extended mind thesis constitutes a strong, indeed a radical version of the theory of the homogeneity of human cognition, for it supposes that the mind by itself changes everything it touches, as it were, and that it reigns absolutely over the cognitive domain as a whole. Whereas Clark imagines that by redefining the extent of the mind in physical space it will be possible to do away with the classical conception of the immaterial

knowing subject due to Descartes, in fact he manages only to rein-
force the methodological solipsism of philosophy of mind and cognitive science by conceiving of the knowing subject as irremediably
confined within an abstract space of individuality.

The Science of Machines

One may nevertheless wonder whether the metaphysical cyborg
argument put forward by the theory of extended mind accurately
describes our relationship to cognitive technologies as a whole. A
recent work by the philosopher Paul Humphreys throws a rather different light on the question.[33] The position Humphreys defends is in
certain respects rather similar to the extended mind thesis, particularly with regard to the importance it attaches to cognitive technological tools in the growth of knowledge, but the burden of his argument
is in fact quite different. "Human abilities," Humphreys says, "are no
longer the ultimate criteria of epistemic success. Science, in extending
its reach far beyond the limits of human abilities, has rendered anthropocentrism anachronistic."[34] Here "anthropocentrism" is to be
understood as referring to the idea that the subjective experience of
knowledge constitutes the archetypal model of all cognition and that
human cognitive abilities constitute the best epistemic criteria available to us. The achievements of contemporary science, Humphreys
maintains, now surpass what human beings are capable of doing and
understanding. This means we must "relinquish the idea that human
epistemic abilities are the ultimate arbiter of scientific knowledge."[35]

As Humphreys rightly observes, there are many fields of science
today—molecular biology, for example, and particularly genetics—
where research is for the most part carried out by machines. To verify
the results of automated inquiry, we consult the instruments that we
have calibrated and the models that describe the functioning of these
machines—models that we refine through trial and error, but that in

many cases we master only imperfectly, for we are incapable of representing all of the interconnections and interactions they comprehend. Furthermore, we are much too slow to be able to check the millions of calculations needed to reach conclusions whose correctness we wish to verify. As a consequence, the processes and procedures that lead to these results remain epistemically opaque to us: we do not know exactly how the machines "know," how they arrive at the results they arrive at. We do have some general idea, of course, because we have a theory that has guided us in devising these instruments. With regard to any particular instance, however, the truth of the matter is that we do not know, to precisely the extent that we are incapable of doing what the machine does; the most we can do is ask it to do the same thing over again.

Nevertheless, it hardly makes sense to say that these essentially automatic, and autonomous, processes, which we are incapable of understanding or thoroughly scrutinizing, are not cognitive processes. They decode genetic information, explore space, analyze financial markets, and construct global climate scenarios. It is not us, it is not an extended and enlarged human mind that obtains these results, but machines and models whose workings we cannot wholly penetrate, epistemically or analytically. Thanks to these machines and these models, present-day science has managed not to extend the mind, as Clark and Chalmers would have it, but to extend the limits of *the cognitive domain* beyond what the human mind can do and understand. There exist cognitive machines capable of doing what we are incapable of doing, which we use to enlarge our knowledge by exploring problem spaces that otherwise would be inaccessible to us.

Many highly complex models—for example, models constructed to forecast climate change—are not analytically transparent. No individual can really understand them in the sense of knowing exactly how they function. The level of complexity at which we can imagine the interactions they generate leaves out an immense quantity of

information and calculation. We calibrate climate models by using information about the past state of the climate to measure a model's success in "predicting" this past state, then modify this or that particular parameter or algorithm until we obtain more satisfactory results. These adjustments are to some extent made blindly, for we cannot always say exactly why the final result is better.[36]

Magnetic resonance imaging (MRI) is another case in point. It is often said that MRI measures blood flow in various parts of the brain. This is not exactly true. What it measures is the relaxation time of the protons of hydrogen atoms in water molecules. The MRI images that play a fundamental diagnostic role in medicine are not perceived visually; they are constructed by the machine on the basis of measurements it makes every one hundred milliseconds. The data collected in this fashion are processed by an elaborate model that rests on a set of assumptions about what the relaxation times of hydrogen protons in water molecules tell us about cerebral activity.[37] The familiar three-dimensional images, color-coded to indicate the most active regions of the brain, are the result of complex sequences of data processing that are not visual in origin. These images are subsequently used by clinicians and researchers to identify regions of the brain that are active in the performance of various mental tasks and to localize lesions and tumors. In place of these images, however, we could print out thousands of pages containing the results of measuring the realignment times of hydrogen protons, for in principle *these data contain all the empirical information needed to construct such images; they contain all the information that has been registered by the machine.* A voluminous printout of this sort would nonetheless be altogether useless. No one, working from such data, can derive and visually display results of any practical use about the patient's brain.[38] The three-dimensional images that play such an important role in clinical medicine and research are possible only on account of the speed and calculating power of modern computers. To generate

images of the brain from these data is beyond what human beings are capable of doing. The human mind cannot by itself transform the data collected by the machine into clinically and scientifically relevant information.

Looking at images produced by a computer or at a computer animation of cellular or molecular motion is in any case very different from looking into a telescope or an optical microscope. It is also very different from consulting an address book. It is no longer a question of observing an aspect of the world that our senses cannot perceive without the aid of an instrument or of remembering an address that we have forgotten. What we "see" is not simply a dimension of the world that until then was hidden from us, but the result of complex cognitive operations. Even when these operations are not a "black box," when they are not epistemically and analytically impenetrable to us, they nonetheless exceed what we are able to do ourselves. Whereas the marks in Otto's address book acquire their meaning from the application of his cognitive abilities, an MRI presents us with information that we are incapable of producing ourselves. It provides us with data we are incapable of integrating and analyzing, in the form of three-dimensional representations of the brain, color images, that we can easily interpret. Such cognitive systems do not extend our mind. They permit us to have knowledge that we could not otherwise obtain. They give us something, in other words, that our own minds cannot supply.

For about forty years now, Humphreys observes, all advances in fundamental physics, molecular biology, neuroscience, genetic engineering, meteorology, and climate research have been made possible either by recourse to computer models that are not analytically transparent[39] or because, owing to the ability of computers to rapidly carry out calculations that would take a human being an entire lifetime to work through, we can give numerical solutions to problems that have no analytic solution. It follows that contemporary science has in large part not been produced by us and, what is more, could not have been produced by us. It is not our mind that has been extended and enlarged, but our knowledge. Henceforth, we are able to venture beyond the limited domain of what our mind can do, because we have created machines, cognitive systems, that are different from what we are. The knowledge that our machines gather in this world beyond our reach can now be displayed in a form that is readily accessible to us—for example, a three-dimensional representation of the brain.

The Mark of the Cognitive?

It is possible, of course, to answer Humphreys by saying that human beings are nonetheless still at the end of the cognitive chain—that the information accumulated by these machines has no meaning and only becomes knowledge, properly speaking, because in the last analysis there is a human mind to make sense of what otherwise would remain mere marks on paper or images on a computer screen. It is therefore indeed the brain's ability to assimilate the functional results of so many different systems that increases our knowledge and expands our mind. The science of machines exists only for us and through us. Without us, machines know nothing.

The question therefore arises: who is doing the knowing here? Who is the cognitive agent? Is it we human beings who know better and more on account of these machines, or is it the machines that know in the first instance and subsequently transmit to us what they have discovered? Can machines truly be said to possess a cognitive faculty of their own, or is it that only some (perhaps all) living organisms are epistemic agents, human beings foremost among them?

The philosopher Mark Rowlands argues in favor of what he calls "epistemic authority" as the relevant criterion for distinguishing processes that are intrinsically and fundamentally cognitive from ones that are only derivatively cognitive. The former operate at the personal level, that of consciousness and the intentional subject; the latter operate at a subpersonal level.[40] We are, as Rowlands points out, hostages to subpersonal processes (the kind, for example, that in David Marr's theory of vision transforms a raw primal sketch into a full primal sketch),[41] not the authors of them. These processes are automatic: they take place without our consent and without our even being aware of them. We exercise no epistemic authority over them; only their results are available to us. We are aware only of what we see, and not of the manner in which our brain constructs visible images. Rowlands holds that these derivative cognitive processes "qualify as cognitive because of the contribution they make to—the way they are integrated into—personal-level cognitive processes to which the authority criterion is applicable."[42] That is, they are not cognitive in themselves, but solely by virtue of their contribution to the mental activity of a conscious epistemic subject.

The cognitive processes implemented by the machines Humphreys considers are in fact similar to the subpersonal cognitive processes that Rowlands describes. They occur automatically and are not conscious to the extent that they are not epistemically penetrable and we have access only to their results. Like subpersonal processes, then, it is uniquely on account of the way in which these cognitive systems are integrated in scientific work—which depends on a set of personal-level processes—that complex models, ultra-high-speed computer calculations, and automated science in general can properly be said to be cognitive, or, if you like, can really be said to be scientific. There is therefore no reason to suppose that these things are not part of the mind in the same way as subpersonal processes, which are likewise epistemically impenetrable.

In this view of the matter, Humphreys's analysis challenges neither the thesis of extended mind—where extension consists in the mind's augmentation by new abilities that are realized in part by processes that take place outside the brain—nor the thesis of cognitive unity, that is, the claim that there exists a unified cognitive domain within which our presumptive superiority, our Cartesian specialness, is preserved. All this seems prima facie all the more plausible as the neural architecture of certain subpersonal processes has often served as a model for the design of cognitive machines.

However obvious and commonsensical it may appear at first sight, the argument advanced by Rowlands nonetheless conceals a number of difficulties. It implies that subpersonal processes, and, by analogy, the automatic processes carried out by machines, are cognitive only to the extent, first, that they include operations of information processing and, second, that they make available to an *intentional subject* information that previously was unavailable to him. While the act of making processed information available to an intentional subject can take place only at the end of a very long series of intermediate operations, it is nonetheless essential that this act occur for the system to be properly said to be cognitive. The availability of information to an intentional subject constitutes what Rowlands calls the "mark of the cognitive."[43]

Now, if a process is cognitive only when it makes information available to an intentional subject, must we therefore conclude that the operations carried out by an automated genome sequencer[44] are cognitive only if and when someone checks the result? The problem is that an automatic system of this sort produces "decisions" that other systems put into effect. Because these ancillary systems act with real consequence on the basis of information processed by the primary systems, it is hard to see why their behavior should not be considered cognitive, even if no intentional agent has had access to the relevant information.

Here is a different example. The Patriot Air and Missile Defense System developed by the United States is entirely automated. It contains a subsystem of radars that are linked to a computer. When the approach of an enemy missile is detected, the computer calculates the trajectory of the incoming missile and an interception trajectory. A Patriot missile is then launched to destroy the incoming missile. All this takes place automatically, without any human intervention. For the moment, at least, the system is supervised by human operators who can cancel the launch of a Patriot missile in the event that the automatic system has made an error—for example, if it has confused a civilian airliner with an enemy missile. The operator's room for maneuver is extremely limited, however: a few seconds at most. Beyond that, it will be impossible to intercept the incoming missile. Response time is crucial. It is the reason why the system was entirely automated in the first place: human beings are too slow. They are incapable of recognizing an incoming missile and calculating an interception trajectory quickly enough.

Do we really want to say that such a system is not a cognitive system? It recognizes an incoming missile, calculates its trajectory, calculates an interception trajectory, and launches a Patriot missile in response. Surely it would be exceedingly odd to say that this system is cognitive only when a human being supervises its operation. The very fact that we depend upon a human being to supervise the system suggests that we are indeed dealing with a cognitive system. The human operator is there to guard against the chance that the system "makes a mistake." Mind you, only cognitive systems make mistakes; mechanical systems break down—but they do not make mistakes!

What Rowlands calls the mark of the cognitive is typical of the anthropocentrism that Humphreys rightly repudiates. In treating human beings as archetypal epistemic agents, it entails the thesis of cognitive homogeneity, for it would have us believe that cognitive systems are all of the same kind, that all other systems resemble human beings. This way of thinking about cognition is proof of the lasting hold that the subjective experience of knowledge has over us. It reflects the naive Cartesianism of cognitive science and philosophy of mind, which have inherited, and accepted without ever really critically examining it, the conceptual structure that Descartes first put in place and that inaugurated the modern era of philosophical reflection about the mind. In this view, first-person knowledge constitutes the paradigmatic form of all cognition, the model of all possible knowledge, and the ultimate criterion of epistemic success.

Automated cognitive systems are distinctive precisely because the information they process typically *does not* become available to an intentional subject. When it does happen to be made available, it is no more than a contingent accident. It is external to the functioning of these systems and in no way essential to their cognitive character, as artificial ethology also very clearly shows. The robotic lobster that recognizes the source of an odor diffused in turbulent water is evidently a cognitive system, and evidently not an intentional agent. It makes far more sense to regard all such systems, not as intellectual and metaphysical cyborgs, but as epistemic actors within a richly diverse cognitive society made up of cognitive systems of several types, some of them natural, others artificial.

Taxi!

All of these things make it likely that Otto will react quite differently than Clark and Chalmers imagine. His memory is no longer very good, and he has forgotten his address book somewhere. So what does he do? He hails a taxi, gets in, and says to the driver: "The Museum of Modern Art, please." By compensating in this way for his defective memory, Otto enhances and enlarges his cognitive *capabilities*— rather than increasing his cognitive *abilities* or extending his mind. The term "abilities" suggests an agent who has augmented his cognitive faculties, whereas "capabilities," understood more or less in the sense intended by Amartya Sen,[45] suggests instead someone who has increased the range of what he is capable of doing, by mobilizing cognitive resources that are available to him but that are not part of him or in any way *his*.

In availing himself of a taxi, Otto quite plainly does not extend his mind. If the metaphor of extended mind seems mildly plausible in the case of the address book, here it has no plausibility whatsoever. The problem has to do with the very expression "his mind." For if the mental processes that take place in the brain of the taxi driver are part of Otto's mind, it is no longer at all clear what "*his* mind" could mean, in relation to either Otto or the taxi driver. By mobilizing the taxi driver's knowledge and cognitive abilities in order to reach his destination, Otto participates in a socially distributed cognitive network that comprises several different types of cognitive systems: the taxi driver himself, of course, but also the GPS on which he relies in deciding what route to take—a system that itself is in constant communication with satellites as well as a great many other automated information systems that increasingly are responsible for managing highway, rail, maritime, and air traffic today.[46] The existence of so vast and varied a network testifies to exactly this, the heterogeneity of the cognitive domain, within which human beings constitute a cognitive system of a particular type, but one that is neither the only possible cognitive system nor the most perfectly cognitive system, much less the indisputable criterion of what counts as cognitive

It is within this network that social robotics aims to introduce cognitive agents of a new type. Substitutes, as we call them, resemble human beings in certain respects more than they resemble computers, or more than computers resemble human beings. Substitutes therefore pose a direct challenge to the metaphor that for fifty years now has dominated our thinking about the human mind: namely, the conception of the mind as a computer. There is nothing in the least nostalgic about this, no lingering desire to affirm the incomparable specialness of the human mind. The new way of thinking about the mind arises instead from advances in human knowledge and technology, which refute the idea of a homogeneous cognitive domain and help us see that we are not the only possible type of epistemic agent.

CHAPTER 3

Mind, Emotions, and Artificial Empathy

[T]he Other is given to me as a concrete and obvious presence that I can in no way derive from myself and that can in no way be placed in doubt or made the object of a phenomenological reduction or of any other *epoché*. . . .

JEAN-PAUL SARTRE

The heterogeneity of the cognitive domain, or, as some may prefer to call it, the diversity of mind, raises the question of whether there is something distinctive about the human mind compared with other types of natural and artificial cognitive systems, and if so, what exactly this peculiarity might be. Philosophy of mind and cognitive science give a contradictory response. On the one hand, they refuse to pose the question, affirming as a matter of principle that we do not have any special quality; we are only cognitive machines, no different from any others. On the other hand, as we have just seen, they hold that we are archetypal epistemic agents, on the ground that human beings uniquely furnish the criterion for deciding what is really cognitive. And yet no sooner has this advantage been conferred upon us than it is taken away, because it is doomed to disappear, they say, once we know how to construct fully conscious artificial agents. In many respects, these agents will also be far more efficient than we are, particularly with regard to memory and the ability to carry out high-speed calculations.

We do not claim to be able to give a complete answer to the question of what the distinctive character of the human mind consists in. Our aim is less ambitious. We wish to show, first, that the Cartesian conception of the mind, which places the knowing subject at the heart of the cognitive process, requires appealing to a social dimension that Descartes himself explicitly rejected. We argue that the Cartesian conception—which instituted what may be called *methodological solipsism,* a principle that by and large has been inherited by modern philosophy of mind, psychology, and cognitive science—rests on a *social cognitive dynamic* whose importance, and even reality, it fails to recognize.

This neglected social aspect is precisely what social robotics sets out to explore. It is also what characterizes substitutes in their capacity as technological devices of a particular type. Later in this chapter, and then throughout Chapter 4, we go on to consider social robotics as *a creative form of inquiry.* This is the second part of our purpose. Social robotics may be said to be *creative* because it looks to build robots of a new kind that are capable of interacting socially with human partners; and this attempt is a *form of inquiry* because it constitutes an investigation into the nature of human sociability. Social robotics is not limited to integrating intelligent machines into an environment that is supposed to be perfectly known and controlled. The machines it designs, on account of both their successes and their failures at social interaction, are also instruments for learning about how human beings relate to one another. Here the basic assumption is that affect is central to what makes us social beings. Robots, in order to have social skills, must be capable of artificial empathy and emotions. From this it follows that affect constitutes a fundamental dimension of mind. But what is affect? This question, which is at the heart of social robotics, will guide us throughout the course of the present chapter and Chapter 4.

Where Is the Mind?

The extended mind thesis seeks to liberate the human mind from the prison of the brain and, in a sense, from anthropocentric prejudice by arguing that the mind can be realized on material supports that are external to the agent. In this respect, it does not dissent from the functionalist thesis of multiple realizability—namely, that cognition is essentially a computational process that can be implemented by very different means: a brain, for example, or a computer.[1] Yet it does not renounce another prejudice, the subjective or individualist perspective, in part because, unlike classical computationalism, it derives from a particular conception of embodied mind in which the question "Where is the mind?" occupies an important place. It puts Otto and, by extension, all individuals at the center of the cognitive process, since they are imagined to constitute its mooring in physical space, and it considers the subjective experience of knowledge as the prototypical form of cognitive activity. Andy Clark could perfectly well have taken the position, of course, that it is when we abandon subjective conscious experience as the criterion of mental reality and accept that unconscious processes are likewise mental processes that the extended mind thesis really makes its power felt.[2] But in advancing a "parity principle" as the ground for determining whether an external process is part of the mind, Clark committed himself to a different view: "If, as we confront a task, a part of the world functions as a process which, were it to go on in the head, we would have no hesitation in accepting as part of the cognitive process, then that part of the world is (for that time) part of the cognitive process."[3]

The brain, or the head, remains the place where the mind resides and from which it extends and expands itself under certain conditions. The brain is the mind's port of registry, as it were, the harbor from which it never strays far enough to lose sight of the jetty. Why should what goes on in the brain enjoy this privilege, as the measure by which the cognitive (or noncognitive) character of processes that take place outside it is to be judged? What could be the reason for this primacy, if not that the cognitive experience of the intentional subject, translated in this case into the materialist language of the brain, is taken to be, as with Rowlands's "mark of the cognitive," the ultimate arbiter of what is "really" cognitive?

The extended mind thesis is proposed as an answer to the question "Where is the mind?" Common sense generally replies "Inside me"—in my head, in my brain. The extended mind thesis adds: not always, for the mind sometimes extends beyond, seeps outside the skull and the skin that covers it. This does manage to qualify the spontaneous response of common sense, but only in a very limited way. An extended arm is able to distance the hand from the chest, for example, because the hand remains attached to the arm. The gesture by which the hand is distanced from the center of the body is possible, in other words, only because the hand continues to be part of the body. Similarly, if Otto's address book extends his mind by helping him to remember, it is because it permits *him*, Otto, to remember the address of the Museum of Modern Art. This is the necessary condition of the address book's being part of a genuinely cognitive process.

Although it is meant to seem paradoxical, in fact this argument does no more than restate the instinctive prejudice that experiencing the world from the first-person point of view reveals the fundamental structure of any cognitive process and therefore of mindfulness itself. Just as it does not recognize the existence of cognitive machines of another type than the human mind, the extended mind thesis, in agreement with the main part of cognitive psychology, presupposes that the subjective experience of knowing, of having one's mind "inside" oneself, constitutes the paradigm of all cognitive activity. Classical cognitivism, which regards the computer as the model of the mind, provides support for this attitude by offering a guarantee of scientific objectivity for what is merely a subjective prejudice.

The impression that the mind is inside of us, incontrovertibly obvious though it may seem in our intellectual culture, is in reality neither primary nor spontaneous. For *experiencing the world* from the first-person point of view is not equivalent to *experiencing the first-*

person point of view. To experience the first-person point of view as a *point of view,* it is necessary to go beyond our direct perception of the world. It is necessary, in other words, to recognize that experience of the world is not the world itself. If it is true that the world is always immediately experienced *from* the first-person point of view, our immediate experience is not *of* the first-person point of view itself, but of being in the world, as one object among others, which presents us with opportunities and dangers in varying degrees. The "discovery" of the mind as something inside us, and, as a consequence, of the world also as being in a certain manner inside us, both representationally and intentionally, rather than our being immediately in the world, is therefore in no way primary or spontaneous. It is secondary and reflexive.

The Mind, Error, and the Other

Discovering that we perceive the world from a particular point of view requires that we have the experience that we are sometimes mistaken about the world and what it contains. Our own mind appears to us only when we discover that our apprehension of the world is not immune to error. A point of view is like a window through which we see the world. As long as the window is perfectly transparent, we do not notice that there is a window. Our cognitive failures, in making the window more or less opaque, give the mind substance by allowing us to appreciate the particularity and the limits of our point of view, which until then, absorbed by the immediate experience of being in the world, we were unable to see. This is a necessary condition—but not, as we will see, a sufficient condition—for our subjective apprehension of the world to interpose itself, in the form of a "mind," between ourselves and the world.

The link between error and the discovery of the mind nonetheless suggests why one finds a long philosophical tradition that, taking the experience of being mistaken as a point of departure, associates the first-person point of view with the subject's privileged access to his own mind.[4] Already in Descartes, an illustrious representative of this tradition, one finds one of the fundamental arguments in favor of this privilege, which has been repeated in various forms up to the present day.[5] The thoughts that my mind contains represent the world, but these "representations" are not false in and of themselves. They can lead me astray only if I ascribe them to the world and if, through an act of judgment, I affirm that they correspond to what it contains, that they *represent* it as it is. In and of themselves, the contents of my consciousness, my mental states insofar as they are here, present in me, as immediate experience, as states of consciousness, cannot be false. I therefore possess a privileged epistemic access to myself, to my own mental states, that is free from error. This epistemic relationship to oneself is superior to any knowledge one may have of the external world, of which it constitutes the necessary condition.[6]

Since Descartes, this reflexive privilege has been accompanied by the view that knowledge of other minds can only be inferred. Whereas I have direct and infallible access to my own mind, I possess only indirect access to that of others. My knowledge of other minds is not only more uncertain than the knowledge I have of myself, it is inferior even to that which I can obtain about the external world. I can directly perceive a tree or a house, but I can only infer, on the basis of their behavior, the existence of other minds. I can never directly perceive the mind of another person, never directly observe his intentions or his emotions.[7] The mind remains concealed inside the fortress of the body, buried in a forest of behaviors, secure against any intrusion from outside—since it can be known only from within. It is therefore not surprising that the existence of other minds, the question of solipsism, should have long constituted, and should still constitute today, a fundamental philosophical problem.

The devaluation of the knowledge of the existence of others' minds, reduced to an oblique and contorted epistemic relationship, an indirect form of knowledge, has the fundamental consequence of shutting the subject up in his relationship to himself. It condemns the subject to seeing others only through the veil of his own mind, with which he alone is in direct contact, an internal theater that contains within it the whole world, but from which others are absent. One can only sense or suspect the presence of others. One can never meet them "in person," can do no more than imagine that these bodies that resemble one's own have a mental life "like mine." As a consequence, the mind is not a part of the world. The mind of another is at best only a theoretical entity; as for my mind, it is not an object in the world, but a stage on which the world can be seen. It is false to suppose that contemporary philosophy of mind and cognitive science have broken with either solipsism or idealism in this connection. Their efforts to embody the mind in the brain of an agent or in material mechanisms merely adopt and recast, as we have seen, this fundamental conceptual structure.

Artificial ethology offers a different image of the animal mind. The cognitive skills of an animal organism cannot be explained by its internal cognitive resources alone. They can be explained only if we take into account, as robotic modeling seeks to do, the complex relationship between its internal resources, its body, and the environment in which the animal acts. Such an approach is consistent with the radical embodiment thesis, according to which the cognitive abilities of a human organism emerge from the relationship between his nervous system and his body in interaction with the environment. The same thesis makes it possible to conceive of a diversity of types of cognitive system, whose characteristics will vary as a function of the particular features of both the terms of this relationship and the relationship itself. The mind may therefore be said to be *radically embodied:* first, because it is inseparable from its particular embodiment, incapable of "escaping" it, of being embodied in a different form (as is supposed in the case of Otto's memory); second, because the mind in this view is not in the world in the way that an object is, but instead as a process. The growth of a plant or an animal, for instance, is not an object but a process, a web of events that take place in the world. In this connection, the term "embodiment" is liable to give rise to confusion, for in the Christian religious tradition it suggests the intrusion or incursion of a divine spirit into human affairs, whereas here the idea is that the mind emerges from purely physical arrangements.

Must we therefore reserve radical embodiment in this materialist sense for animal minds and hold, as Descartes did, that the human mind cannot be explained in so local a manner, through the mere "disposition" of the body's organs, because owing to its far greater universality it supersedes all such contingencies? Or must we say instead that the human mind's particular characteristics are due to its having emerged from a different environment than the one that produced, for example, lobsters and crickets? Descartes's own writings suggest a very different response to these questions than the one that is generally attributed to him and that philosophy of mind assiduously repeats while claiming all the while to reject dualism.

The Evil Demon

Descartes's discovery of the mind, and of the central place it occupies in the process of knowing, proceeded from what he called hyperbolic doubt, bearing upon the whole of the knowledge that until then he had taken to be true. It is important to recall that he considered hyperbolic doubt to be a means of achieving certitude in the sciences, not of discovering the origin of cognition in all its forms. If animals do not have a soul but yet are capable of cognitive behaviors, it follows that the human mind constitutes a particular type of cognitive system, which Descartes admittedly judged superior to all others, apart from divine understanding. His argument is therefore concerned with one type of cognition only, human knowledge, and its conclusions need not apply to other types, still less to all of them.

While recognizing that one is sometimes mistaken is a necessary condition of becoming aware of one's point of view on the world as a particular point of view, it is not sufficient for this point of view to be transformed into a "mind"—that is, into a representation of the world that interposes itself between oneself and the world. This is because, in principle, cognitive errors can be detected and rectified. So far as this can be done, at least most of the time, one's point of view is scarcely apt to acquire the reality, the "ontological weight" needed for it to become something to wonder about.

Thus Aristotle sees a tower that from far away seems to him round. On coming nearer, however, he realizes that it is in fact square.[8] He therefore concludes that the apparent shape of the tower depends on its position in the world and that in changing his own position in relation to it he can discover its true shape. The difficulty facing Descartes in his search for absolute certainty is of another kind. Considering this example, he would conclude instead that he is liable to be mistaken about the shape of the tower at any moment, wherever he may stand in relation to it. Unlike Aristotle, Descartes does not seek to know whether he is mistaken at a specific time, but under what conditions it is possible to be mistaken at all, and what he must do in order *to be able to avoid error at any time*. The problem, as Descartes formulates it, is this: how can one guard against hyperbolic doubt, against the mad suspicion that perhaps there is no tower at all, no world within which such things as towers exist? The philosophical discovery of the self as mind requires exactly this radical sort of uncertainty. But the philosophical fable that leads to this discovery is much more revealing, and much less misleading, than is generally supposed—at least if one is prepared to regard it as a datum, a symptom needing itself to be analyzed, rather than as a true account.

According to Descartes, "I think, therefore I am" offers a certitude that no other thought, belief, or assertion can claim and establishes the priority of the epistemic relationship of a person to himself, to his mind, by comparison with any epistemic relationship to the external world. Now, if "I think, therefore I am" offers greater certitude than, for example, "I scratch my nose, therefore I am," it is because it involves a very particular kind of thought: doubt—the sort of doubt that bears upon everything one thinks one knows. The epistemic priority of the mind is the reverse, the flip side of the claim that any other knowledge is more uncertain and imperfect. It asserts that even when an individual doubts everything, he remains assured of his own existence.

Yet doubting everything is not an easy thing to do. Doubting elementary truths of mathematics, fearing that you are mistaken when you perform a very simple operation—counting the sides of a triangle, for example—is not obvious at all, especially when you wish to convince your readers that you are not so deranged, as Descartes puts it in the *Meditations on First Philosophy* (1641 / 1647), as "those mad people whose brains are so impaired by the strong vapour of black bile that they confidently claim to be kings when they are paupers, that they are dressed up in purple when they are naked, that they have an earthenware head, or that they are a totally hollowed-out shell or are made of glass."[9] To accomplish such a feat, Descartes finds himself obliged to introduce a second character in his fable, usually called the "evil demon" (sometimes "evil genius"). The fact that this demon is an imaginary creature must not distract us from the essential point, that it is a logical necessity.

Let us look once again at the example of the tower and cast it in the form of hyperbolic doubt. As Aristotle approaches the tower, a subtle holographic projection makes it appear square to him, when in fact it is round, as it first appeared to him from a distance and contrary to how it appears to him now that he draws nearer to it. In other words, whereas Aristotle believed that he had corrected his initial error, hyperbolic doubt invites us to suppose that he cannot tell the difference between being mistaken and being no longer mistaken. Consequently, no matter what Aristotle does, whether he stretches out his arm to touch it, whether he walks around it, whether he studies its shadow or observes its reflection in a nearby pond, the tower always appears square to him—when in fact it is round! An evil demon—a cunning computer engineer, as we might think of it today, or a wily film director—does everything in his power to deceive him. Now, what can it really mean in this case to say that the tower "is in fact round" or that someone "is mistaken" when he counts the number of sides of a triangle, when the subject is assumed never to have had the experience of being wrong? Aristotle, by hypothesis, can never conclude that the tower is round since this fact is hidden from him by the demon's artifices; and Descartes, for obvious reasons, will never manage to convince himself that a triangle may have more (or less) than three sides. This situation is utterly different from the one in which Aristotle, approaching the tower, discovers that it is in fact square, whereas from a distance it appeared to him to be round.

What can "be mistaken" mean here, in a situation where one never has, and indeed never can have, the experience of being mistaken because it has been ruled out in advance? What can "be mistaken" mean in a situation where one cannot even conceive, but can only *imagine* the possibility of error? According to the Cartesian fable, it can mean only one thing, namely, that "round" is the way in which the tower appears to another epistemic agent, and that "more or less than three" is the number of sides of a triangle perceived by another, more powerful epistemic agent who may or may not attempt to fool Aristotle or Descartes, to prevent him from discovering how things truly are. The essential thing is not whether this agent seeks to prevent the subject from sharing his point of view, but that subjective error can only be conceived or imagined in relation to another agent having a different point of view. This agent alone makes possible the hyperbolic doubt on which the certainty of one's own existence depends, the assumed epistemic privilege of first-person knowledge of the mind. Hyperbolic doubt requires the existence of another cognitive agent who can take me as an object of knowledge and declare that I am mistaken.

Descartes's hyperbolic doubt and the metaphysical discovery of the mind to which it gives rise require the presence of someone who is able to judge that the subject, Descartes, is mistaken. What permits, though evidently it does not justify, the claim that knowledge of oneself and of one's own mind enjoys epistemic priority over one's knowledge of the world, particularly knowledge of others, is the active presence of a second epistemic agent who exercises his cognitive abilities in the same world in which Descartes exercises his own. For the argument to be successful, for Descartes to be convinced of the certainty of his own existence, Descartes himself must be taken as an object, must be operated on, so to speak, by another epistemic agent who will have "devoted all his energies" to deceiving him. The epistemic priority of first-person knowledge is an illusion, for both the discovery of the mind and the impression that it occupies the center of the cognitive process are made possible by, and flow from, the presence and the action of another epistemic agent who acts upon the subject.

The environment from which the characteristics of the human mind emerge—characteristics that, since Descartes, have been supposed to define its superiority by comparison with other cognitive systems— is a social environment. The cardinal virtue of the Cartesian metaphysical fable is that it implies just this, that *the subjective structure of first-person experience of the mind does not reflect the structure of the cognitive process from which it emerges.* To be sure, Descartes himself drew an altogether different conclusion. Ever since the publication of the *Meditations,* the epistemic priority of knowledge of oneself and the mind has generally been accepted as something obvious. A close reading of what Descartes actually says makes it clear, however, that the discovery of the mind is the result of being fooled—of being fooled, or otherwise led into error, by another epistemic agent. The social environment inhabited by the Cartesian subject is therefore not emotionally neutral. It is shot through with affect, riddled with the anxiety of a subject who imagines not only that he has been led into error by another, but that the error is one that he is not even capable of conceiving.

Methodological solipsism and the priority of self-knowledge over knowledge of others therefore have no sound philosophical basis. There is no reason to take at face value the spontaneous and commonsense answer to the question "Where is the mind?"—neither in its usual version, nor its metaphysical version, nor in the vague and underdetermined version proposed by the extended mind thesis. The mind is neither in the brain, nor in the head, nor outside the agent (much less in Otto's address book), *but in the relations that obtain between epistemic agents.* This is why the question "Where is the mind?"—just like the question "How far does the mind extend?"— scarcely makes any sense at all.

Emotive and Empathic Robots

The branch of robotics that explores the social environment from which mind emerges, a domain largely neglected by philosophy of mind and cognitive science, assigns a central place to the study of affect. Compared with other disciplines, the mixed empirical / theoretical style of research practiced by social robotics gives it a distinctive flavor. On the one hand, it seeks to devise solutions to engineering problems, so that robots will have this or that functional property or

skill; on the other hand, it is guided by a generally shared understanding of sociability and the emotions. There is a tension, however, even a contradiction, between these two aspects. Considering
emotions from the theoretical point of view, social robotics adopts
the methodological solipsism dominant in philosophy of mind, cognitive science, and psychology, whereas in practice, which is to say
for purposes of research and technological applications, affect is seen
to be a social phenomenon.

Social robotics is an extremely dynamic field of interdisciplinary
inquiry situated at the intersection of a number of different research
domains, among them human-robot interaction,[10] affective robotics,[11]
cognitive robotics,[12] developmental robotics,[13] epigenetic robotics,[14]
assistive robotics,[15] and rehabilitation robotics.[16] They encompass a
variety of theoretical frameworks, objectives, and modes of inquiry
that are constantly changing, not only in the case of a particular research project, but even within a particular team of researchers. All of
these perspectives nonetheless converge and contribute to social robotics through their shared interest in a single fundamental question: how can artificial agents of a new type, social robots, be introduced into our network of social interactions? These agents are not
designed to function simply as robotic servants in public settings associated with information, education, health care, therapeutic mediation, entertainment, and so on. The term "social robots" refers more
precisely to artificial agents that are able to work in these fields by
virtue of their ability to *socially* interact with human beings.[17]

The defining characteristic of social robots, as we saw earlier, is
the ability to be perceived by those who interact with them as being
present in the way a person is. To exhibit social presence, in other
words, a robot must give its human partner the "feeling of being in
the company of *someone*,"[18] of being confronted with an "other." This
is not a mere act of projection on the part of the human partner, but
a matter of actually causing him to feel something that is a crucial ele-

ment of face-to-face encounters, the basic relationship from which, in the last analysis, all other social relationships are derived.

Social presence involves something more, and something other, than the ability to decode verbal messages and to respond to them appropriately, or to identify other agents and recognize them in different situations. These high-level cognitive abilities present difficulties in implementation that classical computational artificial intelligence has been working to resolve since its inception. Social robotics studies other kinds of interaction as well, foremost among them the communication that is created by means of what is popularly called body language—posture, gesture, gaze, physical contact. This form of communication also includes affective reactions,[19] in which felt presence is left over as a sort of precipitate. The hope of present-day researchers in social robotics is that by taking into account these various physical and relational elements, which go beyond and at the same time modify the purely computational aspect of the relationship with human partners, it will be possible to find simpler and more satisfactory solutions to problems that have so far proved to be, if not quite intractable, nonetheless very difficult.

The importance of the affective dimension of human-robot relations, especially in helping promote the "social acceptance" of robots,[20] has focused attention on the challenge of building robots that can recognize and correctly interpret the emotional manifestations of their human partners and respond to them appropriately. Indeed, the ability to sustain empathetic relations with human interlocutors is now taken to be the fundamental criterion of successful behavior by robots in social contexts.

The construction of "affective," "emotive," or "empathic" robots, as they are variously called, is currently at the forefront of research in social robotics.[21] Our view is that this is the most promising work now being done, not only in social robotics proper, but also in cognitive science as a whole. The robotic implementation of emotions and empathy has implications that go beyond creating artificial agents endowed with social presence. Apparently straightforward attempts to give emotions and empathy a positive role in robot interactions with human beings inevitably involve a variety of problems at the boundary between cognitive science and philosophy of mind that can then be investigated experimentally. These developments converge with the argument we have developed so far to the extent that social robotics calls into question the fundamental limitation imposed upon cognitive agents by classical cognitive science and philosophy of knowledge: the prison of the solitary mind in which they have been locked away.

In particular, research in social robotics challenges the customary use of singular possessive pronouns in expressions such as "*my* mind" or "*your* mind"[22] and urges us to abandon the assumption, inherited from traditional philosophy, that the mind is essentially something internal, individual, and private, of which the agent is, as it were, the owner. This assumption is common to all the main schools of thought in cognitive science, including, as we have seen, the emerging conception of the mind as a spatial entity, situated in the brain, that may sometimes extend beyond the confines of skull and skin. This view, which purports to be revolutionary for contradicting (or so it imagines) the internalist thesis formulated originally by Descartes, in fact does nothing more than cloak the underlying Cartesian dualism of mind and matter in a fashionable physicalist monism.

The development of a robotics of emotion suggests, to the contrary, that mind is neither immaterial nor something extended in space, but a network—better still, to recall the definition made famous by Gregory Bateson, an "ecology."[23] We have already seen in Chapter 2, and will presently see in greater detail, that mind can be conceived as the coupling or interconnection of different cognitive systems among themselves and with their environment, as well as of the multiple modes of dynamic coordination that link them together on different levels.

Yet the influence of social robotics in both illustrating and helping bring about a paradigm shift in how we think about the mind is not limited to academic debates in cognitive science and philosophy. The way in which emotions and empathy are treated by social robotics has a direct bearing as well on our understanding of the very considerable impact it is likely to have on daily life in our social ecologies. Considering the explosive growth this field of research is expected to undergo in the years ahead, it is scarcely possible to overstate the importance of taking a close look now at what the future holds in store for us.

A Vanishing Divide

Dualism dies hard, and even in social robotics traces of it can still be found. Here two major approaches have long been dominant, one concentrating on the "external" aspects of emotion, the other on its "internal" aspects. These two approaches are associated respectively with the social and individual dimensions of emotion.[24] Research has consequently followed separate paths, one for each domain, even though everyone recognizes that these dimensions are closely related and, what is more, that they have coevolved.[25] But while the distinction between the two approaches is steadfastly upheld within the research community, in practice the difference between them has proved to be unstable and uncertain, and as we will see, many recent developments that are generally supposed to provide support for it have had the effect instead of undermining it further. This state of affairs has given rise to attempts to connect and integrate the internal and external aspects of emotion, as well as the individual and social dimensions to which they correspond. In Chapter 4, we evaluate the success of these efforts. It will be necessary, we believe, to try to go still farther and to reject altogether the dualism underlying this very distinction. Rather than try to join together these two dimensions as separate facets of emotion, we need to conceive of them as integral moments of a continuous dynamic.

The distinction between the two basic orientations of research in social robotics reflects two paramount preoccupations. One involves affective expression by robotic agents, regarded as an external phenomenon; the other is concerned with the production and regulation of emotions in robotic agents, regarded as an internal phenomenon. From the technical point of view, the two orientations assume different forms. Research on the external (or social) aspect of emotions seeks to devise robots that exploit our propensity to anthropomorphism in order to provoke emotive and empathetic reactions in users, whereas research on the internal (or individual) aspects aims at constructing robotic agents whose behavior is influenced by a form of affective regulation inspired by the natural regulation of emotions in human beings and animals.

Accordingly, the distinction between external and internal approaches is associated with two quite different views of robotic emotions. In the first case, the emotions displayed by robots, to the extent that they are *merely* expressed, are thought to be feigned and "false." Exclusive attention to affective expression is bound to produce machines that are limited to simulating emotion—robots that may sometimes be capable of reacting to the feelings of human partners in a suitable and productive fashion but that themselves do not actually feel anything. Such artificial agents are therefore open to the objection that they fool us; that they function by means of deceit, for they do no more than take advantage of the credulity of their human partners, who, moreover, are apt to be vulnerable, as is the case with children with special needs and the elderly.

In the second case, by contrast, the point of equipping robots with a biologically inspired capacity for affective regulation is to create artificial agents having "real" robotic emotions, whose functional role is similar to that of human or animal emotions. The idea, in other words, is to build robots that do not deceive, robots that actually have emotions—"true" emotions, whose expression would be genuine. We are still far from being able to do this, of course. But the basic assumption guiding current research is that artificially reproducing the functional role of emotions is necessary, and perhaps also sufficient, if robots are to be endowed with "real" emotions and that this constitutes an indispensable condition of their entering into authentic affective relationships with human beings.

Beneath these two approaches, and the distinction between the two complementary aspects of emotion that they enforce, we find the same conceptual structure as before: a private mind (in the event, private emotions) shut up inside an agent that can only be known by another mind indirectly, on the basis of the agent's outward behavior. The trick of external robotics—or the lie, as some may think of it—therefore consists in its mimicking these external manifestations without there being anything inside the agent, any inner feeling underlying them. Internal robotics proposes instead to provide the artificial agent with a form of interiority that serves to guarantee the authenticity of the emotions it displays. There is no escape from the dualist schema in either case, however, even if the "interiority" that is in question here, whether present or absent, could not possibly be more materialist. The moral issue connected with the distinction between these two approaches in social robotics, which is inseparable from the question of what constitutes a true and undeceitful affective relationship, is unavoidably affected by the same dualism.

But the habit of associating the external approach with false emotions and the internal approach with true emotions turns out, on closer examination, to be without foundation. Current developments in social robotics are steadily eroding the demarcation between internal and external aspects of emotion, undermining the idea that real, unfeigned emotions—genuine emotions—necessarily arise from a strictly internal process that constitutes the guarantee of authentic affective expression. This challenge to the traditional external / internal division will lead us to reformulate the moral question and approach it from a fresh perspective.

External Robotics, or The Social Dimension of Emotion and Artificial Empathy

Research today on the external or social aspects of emotion focuses on the expressivity that movement, gesture, posture, and proxemic aspects, as well as body type and face shape, give to robots. The purpose of studying the influence of the static or dynamic appearance of robots on interactions with human beings is to equip artificial agents designed to perform a variety of services with forms of expressivity that, by giving emotional color to their actions, will facilitate their acceptance and increase their effectiveness. At the heart of this approach is the idea that empathy plays an essential role in establishing convincing, positive, and lasting affective relationships between robots and humans.

The attempt to build empathic robots is a highly interdisciplinary enterprise, which involves a two-way transmission of knowledge between very different domains, from the performing arts to cognitive psychology and the natural sciences. Basic research is concerned chiefly with the production and recognition of affective expressions, paying particular attention to the factors that favor anthropomorphism, which is to say the spontaneous tendency to attribute beliefs, intentions, desires, emotions—in short, mental states—to animals and to a wide range of artifacts, from stuffed animals to androids. The scientific conception of anthropomorphism has been profoundly transformed and revitalized by recent work in cognitive science and emerging fields of research that come directly under the head of social robotics or that are closely associated with it, such as human-robot interaction and human-computer interaction, with the result that new ways of developing an external robotics of emotion can now be explored.

Psychologists have traditionally conceived of anthropomorphism as the result of confusing the physical and the mental. According to Jean Piaget, this confusion is characteristic of the "egocentric" and "animistic" thinking that children commonly exhibit until the end of the seventh year.[26] More recent studies consider, to the contrary, that anthropomorphism constitutes a fundamental dimension of the human mind. In this view, anthropomorphism is not limited to a particular phase of development; it is *independent of the beliefs of agents as to whether the objects with which they interact have mental properties.* The propensity to attribute mental states to inanimate objects is nonetheless *strongly influenced by the nature and context of the interaction.*[27]

The central hypothesis of this new conception is that our actions take place and our thinking evolves principally by means of dialogue, which creates a context in which we spontaneously treat animals and artifacts as interlocutors. A dialogue context is defined as any communication situation in which turn taking is likely to take place. Such situations can be created in many different ways, through imitation, verbal expression, nonverbal vocal expression (cries, grunts, groans, and so forth), and gestural expression. Currently, there is a consensus among researchers, robustly supported by experimental results, that anthropomorphism derives from the operation of fundamental cognitive structures, which is related to our tendency to think teleologically and to interact through dialogue. The activation of these structures would explain why anthropomorphic projections occur in relationships where the content of the interaction is very poor, and with the full awareness that the interlocutor, typically an animal or an artifact, does not in fact possess the properties that we attribute to it during the interaction. When we plead with a computer not to break down just now ("This *really* isn't a good time, you know!"), we do not suppose that it can hear us or understand us, nor do we imagine that our entreaties will have any effect whatever.

This way of looking at anthropomorphism suggests that feeling comfortable in the presence of an artifact and developing a shared sense of empathy depends on certain minimal conditions that allow it to be treated as an interlocutor in the first place. The external approach seeks to satisfy these conditions through forms of embodiment and autonomous movement that give robots dialogical skills ranging from simulation—gestural and facial reactions suggesting an interactive presence—to actual conversational abilities, leading to the production of suitable verbal responses. The fact that these capabilities can be realized in several different ways makes robots valuable instruments for research not only on anthropomorphism itself, but also on the behavior of artificial agents in various social contexts, whether they are employed as therapeutic aides, teaching assistants, or receptionists, or in any other capacity where their talent for arousing affective and empathetic reactions may prove to be useful.

Throughout this broad field of investigation, in which pure research is conducted alongside the manufacture of technological devices, one of the important sources of inspiration is Masahiro Mori's "uncanny valley" conjecture, which we considered in Chapter 1.[28] To create an impression of familiarity, a robot must resemble human beings in a number of crucial ways, but it must not be *too much like us.* Mori postulated, without really knowing why it should be, that too great a resemblance will give way to a sense of unease, discomfort, anxiety, and, in the extreme case, revulsion. Several recent studies in the field of human-robot interaction have shown that resemblance is not, in fact, decisive in determining how comfortable we feel in the company of robots.[29] In agreement with the revised interpretation of anthropomorphism, these studies indicate that the attribution of emotional and empathic properties to robotic agents depends mainly on the specific characteristics of a given interaction. They agree, too, with the explanation for the sudden collapse of familiarity represented by descent into the uncanny valley that we advanced earlier— an explanation that at once deconstructs Mori's conjecture and accounts for the phenomenon he sought to describe.[30]

All the robots in Gallery 1 fall within the first two categories in the classification of social robots drawn up by Cynthia Breazeal.[31] To begin with, there is a class of *socially evocative robots* that "rely on the human tendency to anthropomorphize and capitalize on feelings evoked when humans nurture, care [for], or are involved [with] their 'creation.'"[32] Next, there is a class of *socially communicative robots*. These offer a natural communication interface to the extent that their capacity for social communication, which rests on a "shallow" model of human social cognition, is nonetheless sufficiently similar to what we are capable of. Breazeal believes that the external approach in social robotics cannot hope to go any further than this and so will be unable to produce artificial agents corresponding to the higher levels of her classification, *socially responsive robots* and *sociable robots*. So long as the external approach fails to draw upon deeper models of human social competence, and, in particular, upon more realistic characterizations of impulses and emotions—internal social objectives, as Breazeal calls them—it will be impossible, she feels, to make robots that are genuinely social partners.

Other like-minded researchers observe that, although the artificial agents of the external robotics of emotion do exhibit true social competence, it is shown "only in reaction to human behavior, relying on humans to attribute mental states and emotions to the robot."[33] This amounts to saying that these agents do not have mental states, that they do not have true emotions, and that they are unable to feel empathy. What passes for social relations with human agents depends on an ability to pretend, to simulate behavior—generally in a benign and fruitful way, but one that nonetheless rests on a mistaken perception on the part of their human partners. Indeed, much work on emotional behavior in robots, some of it significant, "has been driven by the fact that simple behavior may 'fool us' by leading us to perceive robots as having 'emotions' when, in reality, those 'emotions' are not explicitly modeled and just arise as an emergent effect of specific simple patterns of behavior."[34]

Robots can act *as if* they were afraid, for example, or *as if* they were aggressive, but in reality they do not have inner states equivalent to fear or anger, and there is nothing in them, no mechanism or module, that serves to produce emotions. As outside observers, we attribute emotions to robots to explain their behavior; but robots themselves do not have the emotions we attribute to them, for the excellent reason that they do not have inner states corresponding to them, nor do they have any machinery capable of producing them. Robots may well act *as if* they have emotions or *as if* they feel empathy, but they lack the internal dimension required to generate and justify these outward displays of sentiment. One therefore cannot treat the "emotions" expressed by these robots as real phenomena, only as illusory effects that occur as part of an interaction with a human partner, as phenomena that exist only in the eyes of an observer.

In effect, then, the idea that the emotions displayed by robots are not true emotions—that they amount to nothing more than pretending—functions as a methodological principle of the external approach. What is more, it has attracted interest outside cognitive science: not only actors and other performers, but painters and sculptors as well, actively assist and take part in research. This is unsurprising, really, for artists and roboticists share the same purpose: to provoke true emotions in reaction to emotions that, being the products of deliberate pretense, are in some sense false.

External Social Robotics

Keepon

Appearance: Doll-like

Expressive Modality: Noise; body movements:

- lateral movements express pleasure
- vertical movements express excitation
- vibrations express fear

Receptive Modality: Tactile; visual (video camera)

Principal Use: Therapeutic mediator for autistic children; entertainment

Paro

Appearance: Animal-like

Expressive Modality: Cries; movements of body and eyelids

Receptive Modality: Tactile; word recognition; detection of loud noises and of the direction of the source

Principal Use: Therapeutic mediator associated with a variety of conditions: autism, dementia, depression; entertainment

NAO

Appearance: Cartoon-like

Expressive Modality: Movements; posture; gestures; voice; audiovisual and proxemic signals

Receptive Modality: Tactile

Principal Use: Therapeutic mediator for autistic children and treatment of other developmental disabilities; education (teacher, instructor, coach); entertainment

KASPAR

Appearance: Child-like
Expressive Modality: Movements of the head, arms, hands, eyelids; posture; simple gestures; facial expression; voice
Receptive Modality: Tactile
Principal Use: Therapeutic mediator for autistic children

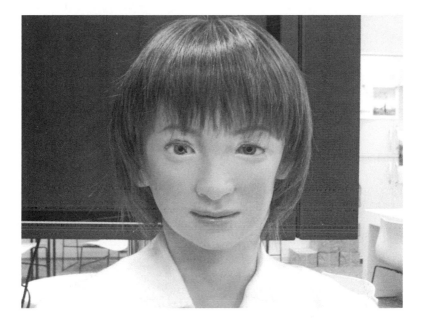

Saya

Appearance: Human-like
Expressive Modality: Realistic facial expression; posture; voice
Receptive Modality: Visual; aural
Principal Use: Education (teacher); receptionist

Face

Appearance: Human-like
Expressive Modality: Realistic affective facial expression (limited to joy, sadness, surprise, anger, disgust, fear)
Receptive Modality: Facial expression; eye movements
Principal Use: Therapeutic mediator for autistic children

Geminoids

Appearance: Human-like

Expressive Modality: Head movements; posture; voice

Receptive Modality: Visual; aural

Principal Use: Research on human-robot interaction; entertainment (theater)

The robots pictured in this gallery, all of them artificial agents developed by external social robotics, are ordered from beginning to end along the Mori curve. The robots that most closely resemble humans are located at the end of the gallery, and those that resemble them least at the beginning. This distribution conceals what, strictly speaking, should be regarded as an ambiguity inherent in the concept of resemblance. The criterion employed here is visual resemblance. According to another criterion, affective expression, the ranking is different. Keepon's abilities, for example, are far superior to those of Saya. Even if visually Keepon resembles us much less than Saya does, it is much more similar to us emotionally. This ambiguity makes it possible to detect the various ways in which roboticists exploit people's instinctive anthropomorphism to draw the user into an affective and empathetic relationship that allows the robot to effectively perform its role of social or therapeutic mediation. The multiplicity of resemblances also lends support to the new view of anthropomorphism as an essential element of human cognitive abilities.

Internal Robotics, or The Private Dimension of Emotion and Artificial Empathy

Research on the internal robotics of emotion is concerned with the role of affect in organizing behavior. In fields such as cognitive robotics, epigenetic robotics, and developmental robotics, the objective is twofold. First, researchers attempt to design and physically implement artificial models of natural affective processes to explore experimentally how emotions contribute to cognition and purposeful activity. The motivation of basic research, in other words, is to arrive at a better understanding of emotion in human and animal behavior. Second, researchers try to use this knowledge to build robots having greater autonomy and an improved capacity for adaptation. Implementation of systems for regulating emotions in robotic agents is meant to equip them with mechanisms that allow them to unilaterally establish an order of priority among various behavioral options. These robots will be capable of selecting appropriate learning and adaptation strategies in response to the needs of the moment, which in turn will increase their capacity for autonomous action with the environment and with other agents, robotic and human.

The growing attention to the internal aspect of emotion is part of a paradigm shift presently taking place in cognitive science, away from the old computational model toward the conception of embodied mind we discussed earlier. This new paradigm, which puts the body back at the center of the scientific study of cognitive processes, has revived interest in phenomena that were neglected by classical cognitive science because they were considered to be merely corporeal, devoid of cognitive value.[35] In robotics, the importance now attached to emotion in cognitive processes—a consequence of the increasingly accepted view that the two are in fact indissociable[36]— has led to the development of radical versions of the embodied mind thesis and, in particular, to an attempt to go beyond the doctrine of sensorimotor functionalism, which seats cognitive architecture in the perception-action cycle. One of the most promising new directions for research, known as organismic embodiment,[37] seeks to recreate within artificial systems the complex interrelations between cognitive processes and affective regulation that are characteristic of natural cognition.

From the methodological point of view, the development of "embodied cognitive science"[38] has led to a gradual abandonment of the sort of representational modeling of cognitive phenomena typically done in classical artificial intelligence,[39] where they are treated as essentially or exclusively computational, in favor of so-called constructive or synthetic modeling.[40] The new approach incorporates hypotheses about cognition in robotic systems and tests them experimentally by analyzing the behavior that these systems express in appropriate environments.[41] Here one of the operative assumptions is that cognition is expressed directly in behavior, which is no longer conceived simply as an indirect manifestation of a system's cognitive capacities.

This constructive approach has a long genealogy in the sciences of self-organization.[42] In its current version, it constitutes a form of explanation that, contrary to the method long dominant in modern science, seeks not to reveal the underlying simplicity of complex phenomena by reducing them to the simple mechanisms that generate them, but to recreate complex phenomena in all their complexity by reproducing the dynamic interrelation of organizational levels: the system, its elements, the environment—and all of their interactions, not least the ones involving the observer.[43]

A paradigmatic example of this approach, applied to the modeling of emotions, may be found in an essay published some thirty years ago by Valentino Braitenberg.[44] The artificial agents imagined by

Braitenberg—vehicles that move around in space, avoiding obstacles, moving away from one another or away from light sources, or moving closer to them, and so on—have a simple architecture that contains only sensorimotor correlations. Yet when these agents interact with the relatively rudimentary environment that their perceptual abilities allow them to recognize, they display behaviors that an observer would spontaneously describe in terms of emotions: they are afraid, they have desires, they are aggressive. This modeling has the advantage of bringing out in a very clear fashion the interactive character of the synthetic approach and the indispensable role of the observer. It has the defect, however, of limiting itself to an abstract sensorimotor functionalism and of applying the synthetic method to no useful purpose.[45] For all its virtues, Braitenberg's thought experiment remains just that, a thought experiment: the cognitive architecture he describes has never been realized, and it lacks any structure or dynamic capable of producing emotions in the way they are produced in natural systems. Affective phenomena in Braitenberg's modeling are open to the same objection as the external robotics of emotion, namely, that they exist solely in the mind of the observer. The synthetic method, as it is used here, is incapable of testing hypotheses about how emotions are produced or of exploring experimentally the role they play in organizing behavior.

Studies on the genesis and functioning of affective processes now being carried out by researchers in cognitive robotics, epigenetic robotics, and developmental robotics under the organismic embodiment paradigm use the synthetic approach in a more productive way. Unlike the classic representational approach of cognitive-affective robotics, which consists simply in using "boxes, standing in for ad hoc mechanisms, that *label* states as 'emotions,' 'feelings,' etc.,"[46] organismic embodiment seeks to use "mechanisms that are argued to be constitutive of representative and / or emotional phenomena,

[an] approach [that] offers greater scope for emergence and flexibility in robot behavioral performance."[47] These mechanisms are implemented in a real robotic architecture, rather than evoked in a purely conceptual fashion in the description of processes supposed to underlie affective behaviors. This makes it possible, in principle, for the observer to wholly recast the dynamics of emotional phenomena within an experimental framework of interactions between a robotic agent and its environment.

It must be emphasized that this research is only in its early stages, and that the promise of the constructive approach remains for the moment unfulfilled. In a striking passage of his recent programmatic manifesto, Domenico Parisi describes the challenge facing researchers today:

> The brain of our robots is extremely simple and it should be made progressively more complex so that its structure and its functioning more closely match what we know about motivations and emotions in the real brain. But the brain is not enough. Motivations and emotions are not *in* the brain. They are the result of the interactions *between* the brain and the rest of the body. The emotional neurons of our robots should influence and be influenced by specific organs and systems inside their body—the equivalent of the heart, gut, lungs, and the endocrine and immune systems. But this is a task for the future. . . . [I]f robots must reproduce the behaviour of animals and human beings and, more specifically, their motivations and emotions, what is needed is also an *internal* robotics.[48]

Parisi's call to arms has not been ignored. Already work is under way aimed at building abstract models of robotic agents.[49] In the meantime, efforts to construct and implement robotic platforms are continuing,[50] and new directions for research are beginning to be explored as well.[51] Unlike attempts in the external robotics of emotion to awaken and exploit our instinctive anthropomorphism, these studies are devoted to creating emotions and empathy in robots. The artificial agents that internal robotics hopes to fabricate will be genuinely empathic, and their emotions real, because authentic. As the cognitive neuroscientist Ralph Adolphs puts it:

> [R]obots could certainly interact socially with humans within a restricted domain (they already do), but . . . correctly attribut[ing] emotions and feelings to them would require that robots are situated in the world and constituted internally in respects that are relevantly similar to humans. In particular, if robotics is to be a science that can actually tell us something new about what emotions are, we need to engineer an internal processing architecture that goes beyond merely fooling humans into judging that the robot has emotions.[52]

The internal robotics of emotion aims therefore at one day producing artificial agents that may be considered emotionally and socially intelligent. The hypothesis guiding this research is that these agents will display an intelligence similar to ours to the extent that their affective and emotional capabilities rest on an internal architecture constructed on the basis of deep models of human social skills (see Table 1).[53]

In seeking to create robots with true, sincere, authentic emotions, the internal approach counts on being able to deflect the charge brought against the artificial agents of external robotics—that their emotions are false, because feigned—by equipping its agents with internal mechanisms that are supposed to play the same role as emotions in regulating human behavior. The principal limitation of this approach is not so much that it reduces emotion to a particular functional role as that, once again, it confines affect to a person's relationship to himself.

TABLE 1

The Internal Robotics of Emotion

Approach	Description
Neuronal Network Model[a]	Based on the dopamine system of the mammalian brain
	Implemented on a simple robotic platform, MONAD
Cognitive-Affective Architecture[b]	Based on three different levels of internal homeostatic regulation and a form of behavioral organization that integrates neural and corporeal activity and sensorimotor activity
	Has different levels of organization that are directly linked to behavioral tasks and to the robot's autonomy
Developmental Affective Robotics[c]	Recursive approach at several levels that takes into account and interconnects the various phases of human physiological and psychological development
	Rests on the principles of cognitive developmental robotics; centered on a knowledge of self and others
	Asserts a parallelism between empathetic development and the development of self / other cognition

a. See W. H. Alexander and O. Sporns, "An embodied model of learning, plasticity, and reward," *Adaptive Behavior* 10, nos. 3–4 (2002): 143–159.

b. See Anthony F. Morse, Robert Lowe, and Tom Ziemke, "Towards an Enactive Cognitive Architecture," in *Proceedings of the International Conference on Cognitive Systems [CogSys 2008]*, Karlsruhe, Germany, April 2008; www.ziemke.org/morse-lowe-ziemke-cogsys2008/.

c. See M. Asada, "Towards artificial empathy: How can artificial empathy follow the developmental pathway of natural empathy?" *International Journal of Social Robotics* 7, no. 1 (2015): 19–33.

Constructing an Affective Loop

The fundamental question for the future of research in the robotics of emotion has to do with the relation between its internal and external aspects. Recent developments suggest that the customary distinction between the two is in fact no longer a settled matter and that social robotics, in upholding the distinction while at the same time disregarding it, harbors a deep ambivalence in this regard.

It upholds the distinction to the extent that the external and internal aspects tend to be treated separately and from different theoretical perspectives. Specialists characterize this divide by means of a series of oppositions, between the social and the individual, the interindividual and the intraindividual, and, above all, between false (or feigned) emotions and true (or authentic) emotions. And yet no justification is given in the current robotics literature for the distinction itself, nor for the series of oppositions that is supposed to correspond to it. Why should the internal aspects of emotion be true and the external aspects false? The question seems never to have been posed. The distinction and corresponding oppositions are simply assumed to go without saying. But in fact they all derive from a commonsensical conception of emotions, central in philosophy at least since the nineteenth century[54] and subsequently taken over not only by classical cognitive science, but also by moderate variants of the embodied mind approach, dominant today.

According to this conception, emotions are internal, essentially private events that take place in an intraindividual space. They are directly accessible to the subject of affective experience and subsequently may or may not be, depending on the case, publicly expressed. When they are, they become accessible to others, but indirectly through the fact of their expression, which remains contingent. Intersubjective knowledge of emotions, just like knowledge of others' minds since Descartes, is therefore never direct. It is the result of rational analysis of the behavior of others, and it rests on an analogy with—which is to say that it is worked out on the basis of—what the subject himself feels and his own reactions in similar circumstances.

It is this instinctive view that leads roboticists to divide their studies into two parts, one bearing on the internal aspects, the other on the external aspects of emotion and empathy—a division that is supposed to correspond to the boundary separating intraindividual space from extraindividual space. In practice, however, it does not really guide research in either one of these two approaches. Each one denies in its own way the distinction that forms the basis for the difference between them; both agree on one crucial point, namely, that affective phenomena occur in a space that encompasses both the intraindividual *and* the extraindividual.

Internal robotics, which claims to be devoted exclusively to modeling the internal machinery of emotion, proceeds, as we have seen, by means of a synthetic or constructive method. Now, this method is inevitably opposed, at least up to a certain point, to the classical thesis that emotions are private events that are generated in intraindividual space and contingently expressed in extraindividual space. The synthetic method in social robotics seeks an explanation of the phenomena it studies in the joint relationship between an agent and the environment in which the agent operates. The relevant unit of analysis in current synthetic modeling is therefore not the isolated individual, but the system constituted by the agent and its milieu—a unit that is inherently relational and, within the framework of social robotics, indisputably characterized as *inter*individual.

The external robotics of emotion likewise, and more obviously, adopts a relational approach to affective phenomena. It, too, violates the distinction between internal and external, only now from the other direction. Even when they are explicitly inspired by the classical view of emotion, studies in external robotics assign the expression of emotion a role that is not limited merely to communicating predefined feelings to a human partner. The affective expression of the artificial agent is conceived and modeled as playing an active part in the genesis of human emotions; in other words, even though the robot's affective expression is not associated with any inner experience or internal regulatory dynamic, just the same it is explicitly intended to produce an emotional response in its human interlocutors. Expression is therefore central to the processes that generate emotions, that is, the affective reactions of human interlocutors, which nonetheless continue to be thought of as inner experiences. Plainly, then, the robot's emotive expression transgresses the strict separation of internal from external, of the individual from the social.

It is plain, too, that the affective reaction of human partners does not rely on any understanding of what the robot feels, whether "by analogy" or in any other way. We know very well that robots feel nothing, and while the physical appearance of Geminoids or of Saya may fool us occasionally, NAO and Keepon are manifestly robots—and this does not prevent them from stirring our own emotions. Even if our emotions are generally not considered to be authentic in this case, because they have been produced by a robot's feigned emotions, which deceive us, the truth of the matter is that this value judgment tries to discredit what it cannot help but confirm, namely, that our emotional response is in fact quite real.

Indeed, in the case of the artificial empathy contrived by external robotics, we are dealing with *pure affective expression*. There is no interiority here, no corresponding inner states. Instead, what we have is a dynamic that does not take place inside, in intraindividual space; it occurs wholly within the space of a *robot-human relational unit*. This interactionist perspective permits us to recognize external robotics as a synthetic approach to modeling emotions and empathy, a minimalist application of the synthetic method that in certain respects resembles the one imagined by Braitenberg's thought experiment.[55] Both rely on a rudimentary cognitive architecture that, in interaction with the environment, allows emotions to be perceived by an observer. By comparison with Braitenberg's synthetic psychology, however, external robotics exhibits two fundamental and closely related characteristics. First, it is more than a mere thought experiment. Second, it creates a real affective and empathetic response in a human being, the robot's interlocutor, who now can no longer be regarded simply as an external observer of an artificial system. The whole point of external robotics consists in just this, that when the interlocutor observes or perceives affective expression in the robot, he immediately enters into the affective process and actively takes part in sustaining the dynamic that drives it.

The boundary between inward and outward aspects of emotions and empathy, which is supposed to coincide with the limits of *intra*-individual space, is continually overstepped as much by internal as by external robotics. They are both agreed, in spite of their divergent orientations, in situating affective processes not in the individual agent, but in the relationship between the agent and his environment, in the *inter*individual unit formed by agent and environment.

It hardly comes as a surprise, then, that an interactionist perspective concerned with reconciling the internal and external approaches should begin to emerge in social robotics today. In seeking to create an "affective loop,"[56] it operates on the assumption that robots must be capable of including the people who deal with them in a dynamic that encompasses the poles of the relationship (robot and human interlocutor), affective expression, and the responses that such expression evokes. The process is interactive, with interaction taking the form of a loop. The essential thing is that the system's operation "affects users, making them respond and, step by step, feel more and more involved with the system."[57] The way to ensure that this will happen is to improve the robot's *social presence*, which in turn means that "social robots must not only convey believable affective expressions but [must] also do so in an intelligent and personalized manner."[58]

Situated at the intersection of internal and external robotics, this interactive approach looks to combine the two dimensions of emotion, both of which are recognized to be necessary to generate a functional affective loop. This requires that the robots be equipped with an internal affective architecture and an external capacity for emotional expression. The degree of complexity of the different internal architectures that may be implemented depends on the way in which they are modeled on natural processes responsible for producing emotions and also on how great a resemblance between natural and artificial processes is thought to be desirable. As in the case of natural systems, expressive abilities are associated with the internal architecture in a variety of ways. A few preliminary examples of robots produced by the interactionist approach are shown in Gallery 2.

First, there are robots like BARTHOC Jr. (a smaller twin of the Bielefeld Anthropomorphic Robot for Human-Oriented Communication) that, according to Breazeal's classification, are situated along the boundary between "social interface" robots and "socially receptive" robots.[59] Even if they are socially passive, they are nevertheless capable of benefiting from human contact. An interactive human-robot interface is implemented through the imitation of the human partner's observed emotions, which gives rise in the robot to a "simple" model of human social competence. Next, there are robots such as Kismet, which Breazeal calls "sociable robots."[60] These artificial agents, no matter that they are still far from having the desired form of intelligence, are nonetheless capable of including their interlocutors in a minimal affective loop. This permits the robot to attain what Breazeal calls internal subjective objectives. In other words, the minimal affective loop that is established between artificial and human agents allows the robot to acquire affective states that are appropriate to the current stage of their evolving relationship and that are determined on the basis of an internal model of social knowledge and emotional intelligence.

The aim is to give robots the means to construct an increasingly stable and efficient affective loop. One way of doing this, currently being studied, is to equip an artificial agent with its own biography, supplemented by a "self-narrating" ability that permits it to provide interlocutors with a "personal" interpretation of its life. The robot therefore has a first-person memory that contains a record of individual experience on which it may draw in communicating with human partners. In this fashion, it is hoped, it will be able to acquire a more consistent and more convincing social presence, similar to the presence of *someone*, to the presence of a person.

The interactionist approach is therefore intended to reduce and eventually eliminate the contrast, and even the distinction, between true human emotions and false robotic emotions. Even if the internal models that produce affective behavior have nothing like the "thickness" of the processes modeled by human or animal physiology, it would no longer be possible to consider robotic emotions as merely feigned or false, for these emotions represent and reveal a genuine inner event. They would no longer be simply a means of fooling us, of making us think that the robot has an emotion it does not really have. Over time, then, it is hoped that the affective expressions of artificial agents, because they are now correlated with an internal dynamics, as in the case of human emotions, will become gradually truer, more authentic.

Note, however, that the attempt to give robots genuine emotions has the unexpected effect of reconciling the interactive approach with the classical conception of the emotions. Interaction, however real it may be, remains external to emotion, which in this conception of an affective loop effectively constitutes an inner state. It is the "inner state" of a robot that tells us the truth about its emotions.

The Interactionist Approach

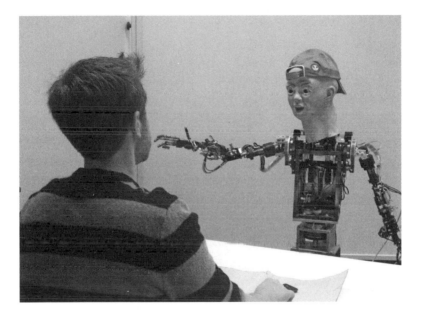

BARTHOC Jr.

Appearance. Partially human-like

Emotional / Empathetic Interaction: The robot recognizes certain emotions (joy, fear, neutral affective state) of its users through the analysis of spoken language. It expresses ("mirrors") these emotions by its facial expressions.

Principal Use: Exploring human-robot interaction

Maggie

Appearance: Cartoon-like

Emotional/Empathetic Interaction: The robot is equipped with an emotional control system whose objective is to maintain the interlocutor's well-being. When it perceives a change in an indicator, from greater well-being up to a certain affective threshold (joy, rage, fear, or sadness), it becomes active; the robot's decision-making system determines the robot's action on the basis of (1) emotional motivations, or drives, (2) self-learning, and (3) its current state. The robot utilizes facial expressions as well as movements of the body, arms, and eyelids.

Principal Use: Exploring human-robot interaction; helping others (as a nurse's aide, for example); entertainment

Sage

Appearance: Cartoon-like

Emotional / Empathetic Interaction: The robot's cognitive architecture is capable of giving it an affective "personality" through changes in its moods: happy, occupied, tired, lonely, frustrated, confused. Certain predefined events gradually lead to these mood changes. The robot's feedback to its interlocutors consists in expressing its moods through speech, the tone and register of voice, and the volume and speed of speech as well as what is said.

Principal Use: Instruction (as a museum guide, for example)

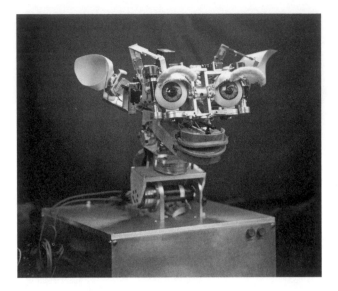

Kismet

Appearance: Cartoon-like

Emotional / Empathetic Interaction: The robot's actions are commanded by an affect-recognition system, an emotion system, a motivation system, and an expression system. The emotion system is inspired by an ethological model for perception, motivation, and behavior; emotions are triggered by events having significant consequences for the robot's "well-being." When an emotion is activated, it leads the robot to enter into contact with something that promotes its well-being and to avoid anything that is contrary to it. The affect-recognition system influences the robot's facial expression; on the basis of these expressions, the human partner can interpret the robot's affective state and modify his behavior and interaction accordingly. The robot's face can express anger, fatigue, fear, disgust, excitement, happiness, interest, sadness, and surprise.

Principal Use: Exploring human-robot interaction; therapeutic mediator for autistic children

To be sure, we are capable of constructing such states, but we will never have direct experience of them, neither more nor less than we will have direct experience of the emotions of another person. In each case, we are supposed to be satisfied with knowing that they are there—that they exist.

The truth of an emotion remains foreign to affective interaction. Evidently, the classical paradigm has not quite yet been abandoned after all.

CHAPTER 4

The Other Otherwise

In equivalence to Watzlawick's statement that "one cannot not communicate" it has been found that also in human-robot interactions one cannot be not emotional.

FRANK HEGEL

Affective Loops and Human-Robot Coordination Mechanisms

The unstable, indeed untenable, character of the distinction between the so-called internal and external aspects of the emotions indicates that the problem of affective behavior must be analyzed in some other fashion. Together with the emergence of the interactive approach, and its paradoxical reversion to the classical view of emotions as private inner events whose expression is secondary and contingent, it suggests an alternative.

Although it lies outside mainstream thinking in philosophy and the sciences, this rival perspective nonetheless has a long history; indeed, one finds an early version of it in Hobbes's analysis of the passions.[1] In recent years, it has acquired important allies in cognitive science, especially within the embodied cognitive science, and particularly in the neuroscientific research on mirror neurons and related mechanisms.[2]

The alternative approach stresses the interactive character of emotions and affective processes and sees affective expression as the core of an intraspecific coordination mechanism in which emotions constitute salient moments. The central hypothesis is that affective expression structures an intersubjective dynamic through which agents mutually determine their emotional states and coordinate their intentions to act. Every one of us participates from birth in this dynamic, which acts at all levels of personal cognitive organization and influences not only our interactions with one another, as individuals and as members of groups, but also our interactions with the environment. More particularly, this affective dynamic shapes our social environment. According to the affective coordination hypothesis, the affective domain in the broadest sense is the site of our primordial, purely biological sociability, which precedes and encompasses the social relations studied by sociology and anthropology.[3] This primordial sociability forms a part of the basis on which cultural, ritual, and social rules of behavior (including kinship rules) are erected and, at the same time, constantly furnishes us with opportunities for deviating from these same rules.

From the point of view of affective coordination, the question of how we acquire knowledge of others' emotions assumes an altogether different cast than the one it has in the classical conception adopted by the computational approach in cognitive science and in the various moderate versions of the embodied mind thesis that are more or less openly accepted in social robotics research today. In the classical view, knowledge of others' emotions, just like that of others' minds in the Cartesian understanding, proceeds indirectly, through inference or by analogy. As a consequence, being able to recognize emotions is crucially important. The affective coordination hypothesis, in contrast, rejects not only the idea that access to others' emotions is obtained by analysis or simulation of expressive behavior, but also that such access rests ultimately on recognition.

It is characteristic of studies of the ability to recognize others' emotions—in the work, for example, of Ekman and Friesen[4] and Izard[5]—that the question of whether one recognizes a particular emotion can be given only one right answer, because recognition is considered to be equivalent to correctly identifying others' emotions. Recognizing affective expression in another person is a matter, then, of determining whether one is dealing with anger, fear, joy, or disgust, for example. To this question there is only one right answer: anger in the case of anger, fear in the case of fear, joy in the case of joy, and so on. Any other response is erroneous and false—evidence of a failure of recognition. When one looks at the matter in terms of affective coordination, on the other hand, it takes on a completely different aspect. I can respond to another's anger by fear, by shame, by anger, or even by laughter. None of these responses, or, if you prefer, none of these reactions, is a priori a false answer. Any of them can permit two agents to coordinate their behavior. Faced with your anger, my shame does not constitute an error, but a coordination "strategy" that may very well lead to an equilibrium.[6] Recognizing assumes a prior act of cognizing—that is, of knowing, in the sense of discovering a particular state of affairs. Reacting is an acting in return, a response, that transforms the meaning and the consequences of the original action. To be successful, a response does not require that a normative criterion concerning the proper interpretation of the original act be satisfied, least of all when a person's reaction acts upon and modifies the affective state of another.

Certain recent results, now securely established in neurophysiology with the discovery and subsequent study of mirror neurons and mirror systems, have given us a good idea of how reciprocal affective action works, even if they do not wholly explain it. Mirror mechanisms produce neuronal coactivation during the course of interaction between someone who acts and someone who observes him. The same neurons are aroused in the observer as the ones that in the observed agent are responsible for the performance of an action or for the display of an affective expression. Vittorio Gallese has proposed that, in the latter case, these mirror phenomena be regarded as embodied mechanisms providing access to the emotions of another person. The access made possible by such mechanisms amounts to a kind of understanding, achieved through a process that Gallese calls "attunement,"[7] which allows the observer to participate in the emotional experience of the observed at an unconscious, subpersonal level by sharing its neurological underpinnings. Gallese considers this form of *embodied* emotional participation to be a basic manifestation of empathy, which he defines as a dynamic interindividual coupling between two or more agents in sensorimotor interaction that suspends the boundary between oneself and another by means of neuronal coactivation. He points to experimental evidence suggesting that the neuronal coactivation produced by mirror mechanisms temporarily prevents the nervous systems of people interacting with one another from being able to tell who is the agent and who is the observer of the affective expression, until sensory feedback takes place. Coactivation, in other words, causes intraindividual space and interindividual space momentarily to coincide.[8]

Our emotions and our empathic reactions are neither private productions nor solitary undertakings. They are joint enterprises in which two or more people take part. Emotions and affect do refer to the body, but this body is not limited to an individual organism; instead, emotions are embodied in what may be called a "social body." From this point of view, the organismic embodiment central to internal robotics, which seeks to unite cognitive processes and emotional bioregulation, meets up with the interindividual embodiment of expression that the external robotics of emotion exploits. In the affective coordination approach, the inner and outer aspects of emotion are not merely juxtaposed, as they are in the affective loop approach described in Chapter 3. They actually combine and merge with each other, for the outer affective expression of one agent is directly responsible for the inner reaction of the other.[9]

Affective expression is a means of direct and reciprocal influence among agents that, even if it is situated at the subpersonal level and supposes a transient erasure of the boundary between self and other, does not cause the identity of the agents in communication with each other to be extinguished. To the contrary, it is owing to affective coordination that the agents' identities are constantly being established, strengthened, and redefined.[10] This influence is direct to the extent that it neither requires nor assumes any intermediary between the one's affective expression and the other's neuronal reaction. The reaction flows solely from the perception of affective expression. It requires no rational calculation or theorizing that is supposed to give one agent indirect access to the inner and private affective states of the other. The influence exerted by affective coordination is reciprocal as well, because the other reacts immediately and expressively to my affective expression, and his reaction brings into existence in my brain a new neural state. Affective coordination thus leads to affective codetermination—an affective loop in the strong sense—that shapes agents' affective expression and their intentions to act.

It is this direct affective influence that the artificial agents designed by external robotics seek to exploit and on which their success as social robots depends. Human beings are caught up from birth in an intersubjective dynamic by which affects are reciprocally determined. Social robots, though they have no mirror mechanisms capable of directly putting them in touch with our emotional experience, are now being introduced into this dynamic. They will need to find a place, a little as our pet animals do, in the affective dialogue that is constantly taking place among us. Unlike our pets, however, social robots have so far found it difficult to fit in. Their limitations arise in large part from the fact that they are not capable of reacting appropriately to the affective expression of human partners. The problem is not just the rudimentary nature of their affective response, but also, and still more significantly, an inability to coordinate their responses with those of their partners.

From this point of view, the ability to artificially produce emotions and empathy does not depend on having a "good" model or a "deep" model of the physiology of natural affects, whether animal or human. It depends instead on the ability of robots to recreate, in the course of interaction with human beings, certain fundamental aspects of the phenomenology of interindividual affective coordination. It depends, in other words, on their ability to include both their human partners and themselves in a recursive interaction process influencing the emotions of human partners and aligning their disposition to act with a corresponding disposition in robots. If robotic agents capable of functioning smoothly as part of this dynamic can be created, it will bring about a social process of human-robot coevolution.

The moral (and political) issues raised by such a process are wholly separate from the question whether robots' emotions are true or authentic. Living with emotive and empathic robots will amount to sharing with them an affective experience that is more or less similar to the one we have in our relationships with pets or that a child has with a stuffed toy animal. These relationships are not artificial. Nor are they false, though they may very well be unbalanced, confused, or perverse. Nevertheless, whether they are altogether healthy, in neither case is it a question of being fooled by our dog or our teddy bear. Childless people who leave all their money to their cats are not really victims of feline cunning. Of course not, it will be conceded—but an intelligent artificial agent can fool us, even if a stuffed animal or a real animal cannot. It is far from clear, by the way, that real animals cannot fool us;[11] but as far as intelligent artificial agents are concerned, the ability to deceive has nothing to do with the presence or absence of an affective dimension. A robot can lead its interlocutors into error by giving them false information, something that it can very easily be programmed to do.

Classically, a concern with the truth or authenticity of an emotion locates the ethical dimension of our affective relationships in interiority and, more precisely, in intention. The sincerity of an intention is considered to be the criterion of an emotion's truth or authenticity.[12] This assumption has no meaning in relation to social robotics, because robots feel nothing. The problem is not that their intentions are insincere, but that they do not have any. To dispose of this difficulty, internal robotics looks to create artificial agents whose behavior is guided by mechanisms that resemble those that are thought to cause, or else in one way or another to accompany, an "inner feeling." By resorting to models of human physiology, in other words, it hopes to be able to make up for the lack of emotion that condemns robots to a life of pretense—to having only feigned and false emotions. The affective coordination approach posits, to the contrary, that the emotions of an agent, whether natural or artificial, are interactive, distributed phenomena, in which two or more interacting agents participate. Accordingly, the ethical dimension is not located in either an internal or an external domain, but in a dynamic that operates on, and transforms, the very agents whose behavior constitutes and sustains it.

Radical Embodiment and the Future of the Social Robotics of Emotion

The idea that the nature of affective relations needs to be reconsidered, and the dynamic of emotions and empathy analyzed in a novel way, has found support in the so-called radical embodiment approach in philosophy of mind and cognitive science. Radical embodiment rejects the extended mind hypothesis, even where its "extension" includes social and intersubjective factors among the external resources on which an agent draws in order to carry out a cognitive task.[13] The radical approach to embodiment proposes a much more revolutionary redefinition of the borders of the mind, freeing it from the spatial dimension within which the debate over extended cognition has confined it up until now.[14]

Radical embodiment, particularly in the enactive version originally developed by Francisco Varela,[15] distinguishes itself not only from the classically dominant tendency in cognitive science, but also from the moderate versions of the embodied mind thesis that seat it in the brain and then extend it, in an ad hoc fashion, outside the intraindividual space. Enaction holds that the mind is situated in—or, rather, arises from—the complex regulative dynamic through which the agent's nervous system couples her body and her environment and thus makes cognition and knowledge of environmental context and of others possible. The radically embodied mind is not a spatial entity in the Cartesian sense of a *res extensa*. It is the result of a dynamic coupling that is irreducible to the classical alternative, inherited from Descartes, between an unextended immaterial substance and extended matter. Because it emerges from a process of reciprocal specification that connects the agent's nervous system with her body and her environment, the mind escapes the spatial framework that any disjunction between internal and external, or between organism and environment, cannot help but assume. The mind emerges, in other words, from a process of coevolution whose imbricated structure inevitably locates mind in the world, and vice versa.[16]

The radical embodiment approach involves more than merely adopting an abstract and speculative theoretical stance at odds with the classical computational conception of a "naked mind" that would be identically implementable in very different materials—as long as a certain "functional equivalence" is preserved between these various "realizations" of the mind. The sheer unreality of the computational conception will be apparent if one considers what a naked mind really implies: take away the body, the environment, and other agents, and all cognitive processes inevitably come screeching to a halt. Modeling and exploring cognitive phenomena become impossible. Contemporary synthetic models of artificial agents, based on the inseparability of brain, body, environment, and other cognitive systems with which agents interact, illustrate the fruitfulness of radical embodiment as a methodological principle. In contrast with a purely hypothetical naked mind, these agents can be designed and actually constructed as recursive systems that perpetually determine and modify one another's state.[17]

Radical embodiment does a better job than the extended mind hypothesis in explaining why Otto would have chosen, as surely he must have done, to get to the Museum of Modern Art by hailing a taxi and asking the driver to take him there. The mind is neither an item of personal property nor something that belongs to isolated individual agents. It is a process in which all agents jointly participate. We believe that extending this principle to the robotic modeling of emotions and empathy will lead to dramatic advances. A relational approach to affective processes, taken together with recent results in the study of mirror mechanisms as well as enactive modeling, will do more than associate the production of emotions with their expression. Not only will a relational approach annul the longstanding divorce of production from expression, it will restore their interdependence by bringing out the complex network of connections that bind together agents through affective processes. In a single stroke, rejoining them does away with all the dichotomies that have been relied on for so long to describe and analyze empathy and emotion: between inner and outer, between private and public, between personal and social, and, not least of all, between true and false.

According to the classical approach, an affective dynamic results from private, internal generative processes that sometimes give rise to an external expression that is both public and social. This expression is supposed to be analyzed subsequently by other agents who are capable of activating in their turn generative processes of the same type, once they have recognized the emotion that has been expressed. According to the affective coordination approach, by contrast, an affective dynamic is a process through which agents mutually transform themselves. They act upon one another on various levels, not only affective and cognitive, but also physiological ("He made me so angry, my stomach hurt").[18] The generation and expression of emotions are complementary and interchangeable moments of the dynamic of affective coordination. They cannot be entirely separated from each other, for affective expression by one person is responsible in part for generating emotion in the other.

Such a perspective deconstructs the true / false opposition by showing the emptiness of the idea that robots will manifest "authentic" emotions—and will not deceive human beings—only when their affective expression proceeds from an internal architecture that reflects, at some suitable level of abstraction, individual animal and / or human affective processes.[19] In effect, then, the relational perspective urges social robotics to explicitly formulate and complete the paradigm shift it is presently undergoing—a shift whose signs we detected earlier, having noticed that research in social robotics constantly and inevitably violates the theoretical distinction between the internal and external aspects of emotions.

The relational perspective proposes a synthetic approach in which creating "robotic emotions" and "robotic empathy" means equipping robots with *human-robot affective coordination mechanisms*. This requires, to begin with, that we no longer look to build robots that artificially reproduce human or animal emotions, conceived as properties of individual agents. We need instead to place the intersubjective dynamic of affective coordination at the center of research on emotion and empathy and develop mechanisms that will make it possible to construct robots that are *essentially interactive affective agents*. These mechanisms, and the emotive and empathic processes they generate, are closely linked with the specific characteristics of the interaction dynamic in which agents are engaged. It is, we believe, on this point—the creation of artificial emotions and empathy as moments of a dynamic of human-robot transformation that coordinates the behavioral dispositions of interacting human and artificial agents, rather than as inner computational or physiological processes that are capable of being given outward expression—that current research has already begun to converge.[20] Social robotics ought therefore to openly declare this to be its aim.

What are the most promising robotic platforms, the architectures best suited to creating artificial agents that will be truly interactive affective agents? Only the coevolution of humans and robots will eventually be able to tell us. Robotic agents of the present generation nonetheless give us a glimpse of what interaction with human beings in one way or another may one day look like. Until now, however, robots have engaged with people on an emotional level only to a very modest extent. Here are three rather well-known robots—Geminoid, Paro, and KASPAR—that illustrate some of the difficulties that attempts to establish a robust human-robot affective dynamic encounter, as well as the limitations of what has been achieved so far.

Geminoid: Social Presence, or Acting at a Distance

Geminoid, the creation of Hiroshi Ishiguro at Osaka University, is an android robot whose appearance almost perfectly reproduces that of its designer. Outwardly, then, Geminoid is Ishiguro's double. Nevertheless, it is not an autonomous robot, capable of acting or moving around by itself. It is a twin, as its name indicates ("Geminoid" is derived from the Latin *geminus*), but only in a very attenuated sense, for its abilities fall far short of those of ordinary mortals, to say nothing of a gifted scientist. Geminoid is essentially a doll, an extraordinarily sophisticated marionette—but a machine all the same, bolted to its chair, attached by a series of cables to a control room, and pumped up by a pneumatic system to look healthy and fit. Geminoid is capable only of moving its head, eyes, mouth, and facial muscles. Nevertheless, it can see and speak. It can also hear what is said to it and carry on a normal conversation. Yet it cannot manage these feats by itself or, more precisely, by itself alone.

Geminoid is remotely controlled, with the help of a computer, by an operator who sees and hears what the robot "sees" and "hears." It is the operator who responds as well. The robot should, in principle, be able not only to transmit the operator's words, but also to reproduce his facial expressions and mouth movements. The operator constitutes the robot's soul, in the altogether classical sense of something that is introduced from outside into an agent's body and that animates it. This arrangement makes it possible for Ishiguro to be traveling, say, in Moscow, and at the same time be present at a meeting of his laboratory staff at ATR on the outskirts of Kyoto.[21] Thanks to Geminoid, Ishiguro can act and react somewhere he is not. In a certain sense, he can be physically present in two places at the same time. As a practical matter, however, Ishiguro's mechanical body is inhabited by whoever—teacher, student, researcher—is sitting at the console in the control room. In addition to making possible what might be called three-dimensional teleconferencing, Geminoid is a tool for exploring the uncanny valley. Endowed with a physical appearance as similar as possible to that of its creator, and remotely controlled by a human operator whose intellectual abilities and capacity for social communication equal or exceed those of an average person, Geminoid ought to allow us to determine more precisely what causes the mysterious uneasiness we feel in the uncanny valley and to have a better understanding of what exactly relations between robots and human beings involve.

Zaven Paré and Ilona Straub conducted a detailed series of communication experiments with Geminoid over the course of three weeks, with the two of them taking turns questioning the robot and operating its controls, Paré sitting in front of the robot while Straub was stationed in the control room, and vice versa.[22] What these experiments demonstrate is the reality of "action at a distance." This kind of action has to do not merely with the fact that Geminoid permits the person commanding it to act in a place where he or she is not physically present. Geminoid makes action at a distance possible in the sense we considered earlier in this chapter with regard to affective coordination, namely, action that takes place at the subpersonal level and that two or more agents take part in.

The experiments are notable chiefly for the robot's stubborn insistence on doing just as it pleases. In reality, Geminoid cannot do much. And owing to bugs, interference, technical difficulties, control problems, and gaps in speech reminiscent of a poorly dubbed film, it has a very hard time doing the few things it is able to do. As a result of all these shortcomings, Geminoid constantly interposes itself between its operator and its interlocutor. Nevertheless, the effect of being there—the robot's social presence—comes through. This effect may be defined as an "action" that the robot exerts on its interlocutor by its presence alone. Anyone who converses with Geminoid has the impression of being in the presence of another person. This ability that it has of acting on us, its human interlocutors, amounts to action at a distance to the extent that it is something that it does to us and that, paradoxically, it does to us by doing nothing, by simply being there.

Knowing that a robot such as Geminoid sees, noticing that it looks, and seeing that *it's looking at me* are not the same things. In the last case, looking at me is something that the robot does to me. "I," the object of its gaze, am sensitive to this gaze. It does something to me. In doing this something, the robot, without moving from where it is, acts on me. It disturbs me, reassures me, worries me.[23] It will be objected, of course, that the robot does not really act on me. On the contrary, it is because I see and understand that the robot is looking at me that I am disturbed. There is therefore no action at a distance here, only an awareness of being the object of another's gaze, of being the target of it, and a knowledge, perhaps innate, of the possible consequences of this state of affairs.

Yet anyone who has even the slightest experience interacting with social robots knows that the robot does not "look" at us in the relevant sense of the term. It shows no interest in us; often, in fact, it sees nothing at all. This in no way changes, or lessens, the impression we have of being the object of another's action. It is true that in this case the interlocutor knew that it was really another person, in the control room, who saw him (or her), but a crucial aspect of these experiments is Geminoid's obstinacy in interposing itself between the two experimenters. Even when the robot's behavior eluded the operator's control, the robot continued to assert its presence, and indeed, in a certain sense, it asserted it still more forcefully.

To be sure, it is possible to say that here the feeling of being disturbed, for example, was unjustified, that it was an error—possibly an inevitable one, Mother Evolution having made us in such a way that we know how to recognize, by certain signs that roboticists exploit, when we are the objects of another's gaze. This ability evolved because being observed by another organism is often a biologically important situation (for example, to be looked upon as a possible sexual partner by another member of one's species)—and sometimes a dangerous one (to be looked upon as prey by a predator).

Yet while knowing that an impression is false changes nothing with regard to the impression itself, as in the case of some optical illusions,[24] we seem no longer to be dealing with knowledge in this case—I do not need to *know*, strictly speaking, that another agent is looking at me—but with something that is rather nearer to a reflex. If this is so, then it is no longer necessary to imagine some intermediary, a mental representation, for example, between another's gaze and my reaction. Even if one assumes the existence of a module, itself representational, that is responsible for detecting the other's gaze, the functioning of this module (by hypothesis, a module in the sense this term has in philosophy of mind) is entirely impenetrable to me. I have access only to its results, which depend on the other's action and which, in this sense, are under its control. Therefore, it is indeed the robot's gaze that acts upon me. In commanding the module's result, it leads to my reaction, typically described as spontaneous, and produces in me a feeling of social presence. The robot acts directly on me from where it is seated no less surely than the physician's reflex hammer does when he taps my knee with it.

The idea of action at a distance seems very strange to us today. Traditionally, it is associated with magic, and since Descartes, the modern scientific view of the world has rejected it. Yet the discovery of mirror neurons explains how such an action at a distance between two agents may be possible. Robotics demystifies it, tames it, makes it intelligible. Geminoid shows that there is no mysticism at work here; acting at a distance corresponds neither to an ineffable sense of there being something "between us," nor to a mere anthropomorphic projection. The presence of another is a phenomenon that can be contrived and implemented with the aid of a machine and that therefore can be reproduced and analyzed. The experiments performed by Paré and Straub cast light on what we find unsettling, disconcerting, and on what makes it difficult for us to recognize this way of acting for what it is. The remote effect of a robot's mere presence, without it having to do anything at all, is yet, paradoxically, an action—an action without either act or actor, as it were! Ordinary language may be confusing here, but we have at our disposal an adequate scientific vocabulary. If this action may reasonably be said to be without either act or action, it is because it takes place and unfolds at a subpersonal level where the difference between self and other is not clearly established. It is nevertheless something that the robot does to *me,* to the extent that *I* passively experience a presence of which it is the cause.

Action at a distance by a robot helps us see that the effect we have upon one another, by the simple fact of our presence, takes place at a subpersonal level. It takes place without our being able to attribute an action to *someone* or to recognize as the author of the action what we discover to be the cause of the experienced effect. The robot, a machine that we ourselves have constructed, does not act by magic. But we are far from knowing exactly how it does what it does, and we are still farther from being able to determine exactly what effect the robot has on us. Unease, discomfort, familiarity, the sense of being kept at a respectful distance—all these feelings are somehow in play, without our being able to locate the effect definitively in one or another register. The exploration of the uncanny valley made possible by Ishiguro's robot is aimed at answering these questions.

Now the essential thing that Geminoid lacks, considered both as a robot, when it escapes the operator's control, and as part of the team that robot and operator normally form, is the ability to react to the presence of another on a level where it is capable of making its presence felt. There are at least two reasons for this. First, Geminoid is a communication interface that, in a sense, is supposed to stay in the background, behind the operator, whom it allows to be present somewhere he or she is not. Second, the operator experiences the presence of his or her interlocutor only as an image on a screen. Unlike the robot's human partner, the operator does not physically experience the social presence of the other person with whom the robot interacts. No affective loop can be established.

Paro, or Proximity: A Return to Animal-Machines

Paro is described by its creator, Takanori Shibata, as a "mental assist robot" designed to interact physically with human beings.[25] It is utilized chiefly as an animal companion for therapeutic purposes in hospitals and homes for the elderly. Paro has the appearance of a baby harp seal (*Pagophilus groenlandicus*) and weighs 2.8 kilos (a little more than six pounds). Like Geminoid, it has no mobility: it is incapable of moving around by itself and has to be carried from one place to another. It is nevertheless much more autonomous than Professor Ishiguro's look-alike. It is not remotely controlled by an operator, and everything it can do it does by itself. Second, Paro is capable of many more distinct behaviors than Geminoid. It can move its flippers and tail. It blinks its eyes, and each eyelid moves independently to give it a more expressive countenance. It can raise its head. It can emit two cries that resemble those of a real baby harp seal. One of these cries is a call; the other functions as a response when it interacts with human beings. Paro recognizes its name. When it is called, it raises its head and turns it in the direction the sound comes from. It reacts in the same fashion to sudden loud noises. Moreover, it can learn a new name if it is given one. It possesses subcutaneous sensors that allow it to detect when it is touched and how gently or roughly it is being handled (whether it is being petted, for example, or tugged at), and it responds accordingly, manifesting through its cries and movements its displeasure or satisfaction. Finally, even if Paro has available to it only a small number of basic behaviors, the number and diversity of the emergent behaviors that it produces in response to being touched are effectively infinite.[26]

Paro is a small, extremely cute animal covered with white hand-made fur. Most people want to touch it or take it in their arms the moment they see it and rapidly become attached to it, mainly because it seems to react so naturally. Since Paro is not mobile, it is always available; there is no need to look for it or call for it to come. Unlike a dog or a cat, it cannot run away; nor is there any risk of its knocking over a glass of water, of breaking a precious vase, or of sharpening its claws on a pedestal table. Furthermore, Paro is robust: it can be handled by many people, even quite forcibly, without breaking.

Paro has much else to recommend it, particularly in a therapeutic setting. Unlike the coat of a real animal, its antiseptic artificial fur transports neither germs nor lice. It is not necessary to housebreak

Paro or to feed it. It has only to be recharged, by plugging one end of a power cord into an electrical outlet and the other end, which resembles a baby's pacifier, in its mouth. Paro is susceptible neither to stress nor depression. It does not become irritable or aggressive when too many people touch it or take it in their arms. In short, it satisfies all (or at least many) of the requirements of animal therapy, with none of the usual inconveniences. Although Paro is used mainly in connection with the care of elderly residents in assisted living facilities and of children in hospitals, it is sometimes, and particularly in Japan, purchased by both individuals and couples as a substitute pet.

Many studies have shown that Paro has a beneficial effect on the mental and physical health of older people. Contact with it improves their cognitive abilities and emotional expressiveness, as well as their ability to manage stress. Moreover, Paro has a positive influence on the number and quality of social interactions in homes for the elderly. It improves the mood of children undergoing long periods of hospitalization, and it reduces depression.[27] Paro provides animal therapy by doing nothing in particular, simply by being there, by being available for repeated encounters—encounters that serve no other purpose or function than social interaction, pure and simple. This is why it is possible to interact with it as one does with a friend, not as one does with a tool or an instrument whose value or utility is associated with a particular task. This may also be why one does not easily grow tired of Paro, why interest does not flag, and why positive therapeutic effects persist, even after several months.[28]

Paro's outstanding quality is probably its physical appearance. We are familiar with baby seals from magazines, television, and films, and we find them adorable. And yet none of us, or almost no one, grew up as a child with a baby seal as a pet. As a consequence, none of us has any particular expectation regarding the behavior of a baby seal. If Paro looked like a dog, or a cat, or a rabbit, we would have a basis for comparison that would allow us to judge whether its behavior is realistic. Moreover, Paro cannot move. A dog, a cat, or a rabbit that cannot move is scarcely credible. The problem is that building a four-legged robot capable of moving with the litheness of an animal poses an immense and, for the moment, insuperable technical challenge. In this regard, Paro's lack of mobility presents a dual advantage. It makes its use in a therapeutic setting for the very young and the very old both easier and more reliable, and it simplifies the technical problem of making Paro seem a credible robotic animal, a plausible "animat." Young seals do not move around much on land (ice floes, actually), and slowly at that.

This attempt to imitate and to faithfully reproduce the appearance of a baby seal therefore leaves a great deal to the imagination, which plays an important role in Paro's success. The robot seems a much more natural, much more realistic animal than it actually is because we have no point of reference. Suspending disbelief is all the easier as no one remembers what it was like having a baby seal as a pet. By accepting Paro as a real animal, we supply a context in which it seems to imitate and reproduce usual forms of behavior. When there is a sudden noise, it looks up, just as an animal or a small child does. It responds to its name. It shows signs of contentment when it is stroked gently and of displeasure when it is handled carelessly. Yet within the comparatively narrow range of what it is capable of doing, Paro's behavior is unpredictable. It reacts differently with different people and differently with the same person at different moments. Thus Paro sometimes gives the impression of acting strangely one day and of being happy another. Variations in its behavior are interpreted as revealing its "moods," its "preferences," and its "personality," even if Paro does not really have any of these things.

Does Paro deceive elderly people who find its company comforting? It is not really attached to them. It will not grieve when they die. And yet no one seems to worry whether a teddy bear deceives the child who sleeps peaceably next to it. In what way is Paro different from a remarkably sophisticated and rather expensive stuffed animal?[29] The answer is not clear.

It is important to keep in mind that there are a great many things that Paro never does, which makes it more like a toy than a real animal. These include all actions involving either inanimate objects or itself, that is, its own body. Paro's behavior consists solely in its relationship to other agents. It never interacts with things. It does not bite balloons; it does not run after balls. Nor does it show the slightest interest in itself. It neither scratches itself nor licks its fur. The only thing Paro does is respond to the voice that calls it or move its tail and its flippers when it is stroked. It never reacts to an event in the world. Even when it moves its head on hearing a loud noise, this does not lead to any action related to the source of the disturbance; the movement has value only as a proof of companionship, letting others know it hears the same noise. Paro is exceedingly social. Indeed, there is nothing in its world apart from the agents who enter into contact with it. It depends entirely on them in order to exist, for it becomes an agent itself only through a relationship to others when it is taken directly as an object of their action. The rest of the time it is a dead thing, an inanimate object.

Paro has no interests of its own. It is sensitive only to the interest shown it—whence the impression that it is interested in you and no one else. While it is true that Paro reacts when it hears its name, for the most part interest must be communicated tactilely for it to react: Paro needs to be touched, taken in one's arms, carried. By their very nature, intimate encounters of this sort tend to exclude others. If Paro is exceedingly social, then, it is nonetheless social in an exclusively individualistic way. It seems to establish a special relationship with every person it meets. What is more, it will never betray to anyone the secrets that you have confided in it.

At first sight, there is something paradoxical about this exclusive individualism, for Paro often facilitates social interaction among the residents of homes for the elderly. People who no longer speak to one another suddenly begin spending more time together, conversing in common rooms.[30] Yet if Paro facilitates social interaction, it is by coming between people, as an object rather than as an agent. Paro can take part only in an extremely limited way in the conversation of a group for which it is the center of attention. In this case, it is hardly an interlocutor—more a facilitator or a pretext. It is *something that is spoken of,* not *something that is spoken to*—unlike what happens in its one-to-one relationships. Paro facilitates conversation and social contact among those gathered around it, without being a member of their circle. It gives them a subject of conversation, something that is agreeable to talk about and show an interest in, but it cannot itself converse with them. In its one-on-one relationships, Paro functions more as an agent, though here again it remains apart from, and indifferent to, the activities of the individuals whom it provides with a reason for coming together as a group. It has no interest in anything that is said *about* it. Only someone who speaks to it directly can attract and hold its attention. Paro is incapable of entering into a relationship with others in the way that its availability allows them to interact among themselves—which is to say, through the intermediary of something that brings them together by separating them. No object can mediate a relationship between Paro and others; no person can mediate a relationship between Paro and objects.

KASPAR and Caring

KASPAR (an acronym for Kinesics and Synchronisation in Personal Assistant Robotics) is a robot originally conceived as part of a research project begun in the late 1990s by Kerstin Dautenhahn and her collaborators at the University of Reading in England.[31] Initially, the objective was to develop "robotic therapy games" to facilitate communication with autistic children and to help them interact with others.[32] In 2005, now at the University of Hertfordshire, the KASPAR Project was formally launched with the aim of developing a "social" robot having two missions: first, and mainly, to be a "social mediator" responsible for facilitating communication between autistic children and the people with whom they are in daily contact, other children (autistic or not), therapists, teachers, and parents, and also to serve as a therapeutic and learning tool designed to stimulate social development in these children. Here we have a complex objective that involves teaching young autistics a variety of skills that most of us master, more or less fully, without any need of special education: understanding others' emotions and reacting appropriately to affective expression, expressing our own emotions, playing with others while letting everyone take turns, imitating others, and cooperating with others. The idea of using playmate robots for therapeutic purposes came from a well-attested observation in the literature on autistic children: early intervention can help them acquire cognitive and social skills they would otherwise be incapable of developing.

This therapeutic and educational project requires a robotic partner whose social presence is at once obvious and reassuring, because its behavior is easily anticipated and understood. KASPAR is a humanoid robot the size of a small child, about three years old. Its physical appearance, in keeping with the usual interpretation of the Mori effect, is not overly realistic. Dautenhahn and her team achieved this reduction in the complexity of social communication, while also managing to avoid the complications created by excessive likeness, through an extreme simplification of facial features. The robot's face is a skin-colored silicon mask devoid of the details that normally make it possible to determine age, gender, emotional intensity, and so forth. On the one hand, this deliberate lack of definition gives free rein to the child's imagination, allowing him to think of KASPAR as a playmate, or at least as someone he feels comfortable being with. On the other hand, it gives KASPAR's designers considerable leeway in engineering (and later programming) customized versions to suit a variety of needs.

Autistic children interact quite readily with KASPAR from the first meeting.[33] In its present version, the robot is dressed as a little boy. It is capable of moving its torso, arms, and head. It can also open and close its mouth and eyes. This restricted range of movements gives KASPAR a "minimal" emotional expressiveness, uncomplicated and easy to interpret. Taken together, the movement of eyes and arms, posture, and voice permit it to express several basic emotions: joy, sadness, surprise.

KASPAR is not an autonomous robot. An operator controls its movements and speech using what is known as the "Wizard of Oz" technique.[34] Its usefulness as a social mediator and as an instrument of therapy therefore depends on human intervention. KASPAR succeeds in getting autistic children to take part in a wide range of interactive games that usually are not accessible to them because they involve activities such as imitation, turn taking, and shared gaze. With the aid of different game scenarios, Dautenhahn's team set out to evaluate KASPAR's educational and therapeutic potential for treating autism in children, in addition to developing their social skills.[35]

KASPAR's principal role is as a social mediator in the relationship between an autistic child and a therapist, a teacher, or other children. In this regard, it is often used to teach autistic children to express their own emotions (sadness or joy, for example) and to recognize affective expression in others. Studies suggest that KASPAR's expressive minimalism, implying an extreme simplicity of interpretation, furnishes these children with a sufficiently predictable and reassuring social context that they are able to play with others and to try doing new things. Another application of the same type, made possible by the robot's epidermal covering, RoboSkin,[36] involves a game that teaches the child to exert an appropriate degree of force when inter-action brings him into physical contact with others. Some autistic children who exhibit tactile hypersensitivity or hyposensitivity find it difficult to properly modulate the strength they bring to bear while playing. Here again, interaction with KASPAR furnishes such children with a protected environment that is easily understood and reassuring. When a child fails to correctly judge the amount of force that should be used, the interaction nonetheless continues without interruption and the child is not made to feel rejected. Instead, KASPAR sends a clear message—"Ouch! That hurts!"—without getting angry and ejecting the child from the game, as other children often do.

KASPAR is often used with so-called high-functioning autistic children as well. To be able to join in playing a video game, for ex-ample, the autistic child must imitate the movements that the robot executes. In this way, thanks to KASPAR's mediation, he learns to co-operate and to take turns playing the game with a normal child. The robot can also be used to help autistic children discover their body image, following its example. By touching and naming this or that part of its humanoid body—the nose, the ear, the arm—the robot teaches the child to do the same thing.

The effectiveness of all these learning experiences proceeds from the fact that the robot provides the child with a relaxed atmosphere in which, unlike the social situations he is used to, he is protected from the awkward and often distressing consequences of his many errors of interpretation. The robot never reacts in a reproving or dis-missive way when the child behaves inappropriately. Instead, it gently corrects him while at the same time providing reassurance, by means of firm and unsurprising responses that are unlikely to be misunder-stood and that encourage the child to persevere in the difficult work of learning social skills. The ability to give both comfort and cheer—

feelings we experience in the company of pets as well—is common to many robots used in a therapeutic setting, not only ones such as Paro but also substitutes designed to help people who have suffered a stroke or other cerebrovascular accident, or any injury that suddenly compromises one's ability to perform elementary daily tasks. In this connection, where motor and cognitive rehabilitation is inseparable from *social rehabilitation*, several studies show that many patients prefer to perform therapeutic exercises under the supervision of a robot rather than a human nurse, whose presence is apt to disturb or embarrass them in trying moments, when they prefer not to have to compare themselves to other people.[37]

The reassuring character of such interaction is also the basis for a number of recent developments, among them the suggestion that KASPAR might be used by police for questioning normal children who are victims of abuse or surviving witnesses of accidents or criminal acts. These children often find it difficult, or are afraid, to speak candidly to an adult. According to the Metropolitan Police in London, they frequently do not furnish useful leads, and they sometimes give false or misleading information. The guiding assumption is that when interrogators, even if they are trained social workers, hear certain accounts of abuse, they find it very hard not to transmit nonverbal signals that the child finds disconcerting, causing him to hesitate and either stop talking or change his story. A robot, it is thought, may be perceived by the child as a more neutral and less threatening conversational partner, making it possible for him to treat a delicate situation as a sort of game.

This proposal nonetheless raises important ethical issues. The child is not aware that he will be speaking to a human being when he is made to interact with the robot, and the interaction is for the express purpose of inducing him to give information that he might not give otherwise. In this sense, it is indeed a question of fooling the child, of making him believe that he is speaking to a robot when in fact a human being is listening in and doing the talking. This deception nonetheless assumes a form opposite to the one for which robots are usually reproached. Usually, a robot is accused of having false emotions, because it pretends to have emotions like those of human beings, whereas in fact, it is claimed, they are not the same, because there is no corresponding internal state. Here, to the contrary, the idea is to reassure the child by making him believe that he is dealing simply with a robot. If the child is fooled, then, it is not by a robot, not by a machine, but by adults, by other human beings.

KASPAR's undoubted successes in other domains expose the shortcomings of most current ethical thinking about the use of robots in educational institutions, hospitals, and specialized centers for children and adults with particular disabilities. Social robots are often thought of as a way of delegating the obligations associated with such care to machines, rather than taking responsibility for it ourselves. But robotic interaction partners such as KASPAR cannot *replace* human beings. They can only *support* them in providing aid and treatment. In this sense, robots are *substitutes* for human partners—intrinsically *temporary* substitutes, designed to make it easier to establish social ties with those who, for one reason or another, have a hard time fitting into their social environment. What Phie Ambo in her film calls "mechanical love"—what we call artificial empathy—cannot simply be seen as a matter of false emotions, of emotions that have a positive effect only by manipulating and deceiving robots' human partners. The artificial empathy and other emotions that emerge and develop in the course of interaction between robots and human beings give rise to an affective dynamic that produces appropriate social responses. This dynamic makes it easier for those who have trouble interacting with others to enter into a social ecology—a human world that otherwise would, for the most part, be closed to them.[38]

The Other Otherwise

Geminoid and Paro suffer from the same weakness, but in different ways. They cannot interact with objects, only with other agents. To be sure, Geminoid can speak and hear. Unlike Paro, it knows when it is being referred to, and it has no shortage of excuses for conversation, even if in truth it is not the one who is actually speaking and hearing. All that Geminoid can do is accomplished entirely within the bounds of conversation. Like Paro, it has no relation to the world apart from its human partners, who together make up the whole of its existence. Both Geminoid and Paro may therefore be said to suffer from an excess of sociability for its own sake.

As human beings, very few of our social relations are direct. They are almost always mediated either by a material object on which our action bears, separately or jointly, or by an activity that functions as a pretext (having a drink, taking a walk, and so on) and that itself is mediated by conversation. This latter kind of mediation is plainly inaccessible to Paro: no activity can interpose itself between it and its partner as a pretext, even if Paro itself can play this role with others and become the pretext, the occasion of their meeting. Mediation of this kind is also to a large extent impossible for Geminoid, despite its capacity for speech, for it has no interest in its immediate surroundings, in the world in which it finds itself here and now.

The expression "has no interest" must not be understood here merely as a subjective disposition, but as an objective relationship to one's environment. In this sense, we are interested in the world to the extent that it acts upon us. Now, the world that acts upon Geminoid is not the world common to the robot and its interlocutor, but the system that connects Geminoid to its operator. This is why the common world cannot mediate the robot's relationship to its interlocutor. It is also why, even if the robot makes its presence felt by its interlocutor, the interlocutor in no way makes his presence felt by the robot. While the robot acts at a distance, on me, I do nothing to it, unless I touch it directly. And yet, as Paré and Straub's experiments show, Geminoid's reactions to my direct action, unlike those of Paro and KASPAR, are unrelated to the interaction taking place. They have no communicational value.[39]

If Geminoid manages to do something that ordinary teleconferencing cannot, namely, make a three-dimensional physical and social presence felt, its human interlocutor, as a result of the robot's indifference to him and to the physical environment they share, sees his own presence reduced to a meager sampling of images and words that are seen and heard elsewhere. Every exchange, every communication with Geminoid remains a prisoner of language, a domain from which Paro, by contrast, is barred. And yet Geminoid stands mute at the gates of physical communication, a domain in which Paro enjoys a wonderful fluency. As a mere means of communication, Geminoid manages neither to make itself entirely disappear behind its operator, for whom it acts as a spokesman, nor to step forward as a genuine conversational partner. The principal reason for this failure is an absence of reciprocity at the level of affective engagement. There is no response from Geminoid to the reaction that its own action, performed at a distance, provokes in me. I am nothing to it. I exist only for the person who commands it from the control room. Geminoid is capable only of a simple, imperious relationship: asserting its own presence. It is incapable of reacting to the presence of another in the same way that it makes its own presence felt.

Paro, by contrast, acts and exists as an agent only through another agent, a human partner, who makes it the object of his attention. In reacting to this attention, Paro arouses a response in its partner, which is manifested by an action that in turn provokes a new reaction in the robot. This is why interaction with Paro establishes a much more complex affective dynamic than anything that is possible with Geminoid. Paro lacks the ability to withdraw from the interaction in which it is engaged and take an interest in its surroundings, whereas Geminoid, owing to its ability to speak, inevitably invites people to interact with it. Paro, being incapable of speech, remains the captive of a purely social relationship from which there is no escape, no opening onto a wider world.

Paro and Geminoid both suffer from a severe deficit. Although they are present in this wider world, they cannot take an interest in it. Their behavior is exclusively social. KASPAR has no such handicap. Its educational function, which rests in large part on imitation, allows it to use the interest the child shows in it as a way of getting him to show a new interest in the world and in himself. For KASPAR to be able to play this role, its world cannot be reduced simply to the interaction that is taking place at a given moment, as is the case with Paro. Like Geminoid, KASPAR is a semiautonomous robot, remotely controlled by an operator, which in principle gives it access to a body of specialized educational knowledge. Nevertheless, like Paro and unlike Geminoid, it is capable of reacting to direct contact; of reacting to the child on the same level as he acts on it, which permits it to modulate the interaction while it is taking place.

As imperfect as they may be, these robots give us at least a glimpse of what an ideal artificial social agent might be like. Paro and Geminoid, because they lack the ability to act in any true or deep sense, are unable to relate a capacity for action to a process of affective coordination with human partners. KASPAR, on the other hand, is able to do this, and it uses this ability to socialize the autistic child by including him or her in a process of affective coordination. KASPAR's limitations are due to its lack of mobility and to the fact that it is a semiautonomous robot. Just like the exoskeletons one encounters in Japanese manga, this artificial agent is an agent only because it is not in fact wholly artificial, being "piloted" by a human being. Hence the question arises: with what does the autistic child who interacts with KASPAR interact? Whatever reply may be given to this question, are we obliged to suppose that KASPAR, which helps the child to develop skills that very probably he would not be able to acquire otherwise, deceives him?

Would genuine artificial agents, fully autonomous social robots, be "mindless" machines because they do not possess mental states similar to ours? Would they be unfeeling, incapable of empathy? We think not. Our view is rather that machines that are capable of assisting in the coordination and continuous cospecification of human social relations will integrate themselves into a complex ecology of mind, each in its own distinctive way. Robots capable of establishing emotional and affective interindividual coordination processes with human beings will themselves be without private, inner feelings because there are in fact no truly private, no truly inner feelings, contrary to what we have long been accustomed to suppose, as a matter of common sense, and what current research in robotics continues to imagine. If this is so, a new set of questions will need to be addressed. Could social robots really help and care for us? Could they really sustain relations characterized by trust, friendship, or even love? If they could, what would that mean? Would these relationships be authentic?

They would certainly not be authentic if by "authentic" it is meant that they would be identical to the ones based on benevolence, trust, friendship, and love that we form among ourselves, for inevitably they will be different. But that does not mean that they will necessarily be in any way false or misleading. Affective relations among human beings, no matter how authentic they may be, are often misleading and deceptive. There is no reason to believe that artificial agents capable of affective and emotional coordination will not produce their own fair share of frustration and disappointment.[40] Human-robot coevolution may be expected to give rise to new kinds of relationships that will have their own characteristics and their own phenomenology. Here, no doubt, just as in our relationships with one another and with our animal friends, human beings—and perhaps artificial agents as well—will sometimes wonder whether the emotions and the empathy displayed by their partners are authentic, whether they are real.

If the prospect of coevolution between humans and robots inevitably raises many ethical and political questions, it also opens up new paths of inquiry. One must resist the temptation to ignore these questions, or to close off these paths, without first thinking very carefully about where they lead. There is a danger in treating human-robot coevolution as just another transformation, similar to the ones now under way (or anticipated soon to occur) in biomedical, information, and communications technologies, particularly with regard to social media; or to the ones accompanying the advent of powerful automated systems for managing transportation networks, financial markets, banking transactions, and surveillance; or in vitro fertilization techniques and organ transplants; or transgenic plants, animals, and, soon, human beings (children who have three biological parents are already a reality). It is clear that we are living in an age of unparalleled upheaval. Technological advances have led to profound moral, political, and social changes, the extent of which no one today is able to determine exactly. The difficulty comes, on the one hand, from the fact that what we commonly call progress is much less evenly distributed than imagined by either those who warn us against the coming of the apocalypse or those who gladly welcome the dawning of a radiant future. On the other hand, and in an opposite fashion, the difficulty of estimating the scope of the social transformations associated with technological advance comes from the fact that these transformations are much more coherent and, in their way, more uniform than supposed by the many regional ethics[41] that seek to regulate them. The cause of our myopia, in other words, is that we tend to exaggerate both the global effects and the disjointedness of the phenomenon.

The developments we examine in this book involve technological objects of an entirely novel type. Social robots have the power to bring about a metamorphosis in the ecology of our social relations that is altogether different from the ones that other technological objects are now bringing about. Hannah Arendt distinguishes between three types of activity associated with three fundamental aspects of the human condition: labor, which is made possible by life itself, by our biological existence; work, an expression of the "unnaturalness" of human life, which is to say the production of a cultural world filled with material and intellectual artifacts; and finally action, or politics, which corresponds to the "human condition of plurality" as a conse-quence of the fact that the world is inhabited by men and women of different races, not by a unique, singular being called "Man."[42] Most, if not all, of the technological objects that human beings have fabricated up to now—including the most recent ones—have both renewed and utterly transformed either the biological condition or the cultural condition of humanity. The creation of what we call sub-stitutes holds out the prospect of enriching and transforming the plural condition of humanity. It promises to introduce new creatures among us—creatures that, after the fashion of animals, only in a dif-ferent way, will be at once like us and unlike us.

CHAPTER 5

From Moral and Lethal Machines to Synthetic Ethics

The idea of a device with unerring judgment about what's really going on, one immune to all stress and distraction, is as appealing as ever.

JAMES R. CHILES

There are many artificial agents that have little in common with the robotic substitutes we have studied so far, especially in Chapters 3 and 4. It will be necessary now to show more precisely how they differ and why the ethical and political problems they pose differ from the ones raised by the introduction of substitutes in a variety of social settings.[1] This is all the more important because, unlike substitutes, which for the moment typically exist in the form of prototypes and scientific instruments having specialized uses, mostly in health care and related fields, the artificial agents to which we now turn are already among us.

They are numerous, influential, and to a large extent invisible since in many cases they are not individual machines, discrete objects that can be detected in the physical space we inhabit. Nevertheless, they act; they are agents. Paradoxically, however, it is generally impossible for the user to identify the author of the action that they perform. They may be described as "partial" or "integrated" agents or else as "analytic" agents, depending on the level at which their function is analyzed. Such an agent is partial or integrated to the degree that, detached from the complex technological system of which it is a component, it is incapable by itself of doing anything whatever. Even so, like the banking algorithm that processes your request for funds, analytically it must be said to be an agent. In the case of a banking algorithm, it corresponds to a few lines of code that "decide" whether to approve or reject your request. Such an agent is in no way an actor, however.

Other artificial agents, like drones and similarly autonomous vehicles, are genuine robots. They are three-dimensional objects in physical space. Yet here again, most of them amount to nothing more than executive agents. Unlike substitutes, which are active partners in conversation or some form of emotive exchange, these robots, like partial or integrated agents, are agents whose whole function consists in making a decision, performing an action, or carrying out a particular operation.[2]

Current ethical thinking about intelligent machines focuses on these two fundamental characteristics, exhibited by the majority of artificial agents that are among us today: they function as integrated components, and they are dedicated to a single purpose. Substitutes, in contrast, because they are explicitly designed to be fully social actors, pose different types of problems and invite us to think about their ethical implications in a different fashion. From this point of view, our relations with substitutes cannot be considered to be merely a matter of controlling the behavior of agents having no moral autonomy. They cannot be limited to laying down rules that restrict their freedom of maneuver and constrain the operation of the complex systems in which they are embedded, in something like the way that banking and financial markets are regulated today. Partial and other executive agents are essentially machines that work for us. There the important thing is to limit the consequences—sometimes unforeseeable consequences—of their actions, as we already do in the case of simpler machines. Substitutes are indeed machines, but they are also individual agents, a new and different type of social actor with which we now increasingly find ourselves having to engage. As a result, they are, or at least they may soon become, instruments for research in the ethical domain as well. Just as in some Japanese manga, they present opportunities for moral discovery and learning. Preliminary investigation of the open-ended behavior of substitutes such as KASPAR has already given us a glimpse of what this might entail.

Robot Ethics

The last few years have seen the publication of books with titles such as *Robot Ethics*[3] and *Moral Machines*,[4] as well as quite a few articles dealing with particular aspects of the subject. The current fascination with robot ethics grows out of a long-standing interest in so-called machine ethics, which bears on the relations between human beings and computers.[5] Machine ethics for the most part resembles other domain-specific ethics (or regional ethics, as we call them), such as medical ethics or business ethics, whose aim is to formulate acceptable rules of conduct in various fields of endeavor. There is nonetheless one essential difference, which we will come back to in due course—namely, that here the rules of appropriate behavior are incorporated directly in a computer program that marks out the boundaries within which the computer may determine what response is to be given to a particular human action or request. A major technical hurdle is ensuring that the responses given by a computer or computerized system will be sufficiently polite and respectful to be readily accepted by human beings. This is why machine ethics seldom ventures beyond the limits of good manners and sound business practices and provides little more than a professional code of conduct for computer scientists.[6] The questions it asks are important, but they fall not so much within the traditional framework of ethical inquiry as within a style of thinking that has become familiar in the past thirty years or so with the multiplication of regional ethics.

Robot ethics[7] sets forth a much more ambitious and revolutionary research program. It is concerned with making robots "moral machines" by teaching them the difference between right and wrong.[8] The avowed purpose is to devise ethical rules for artificial agents— that is, to create artificial moral agents (AMAs in the conventional shorthand). It therefore comprises both a technological dimension— constructing artificial agents that are sufficiently autonomous to be genuine moral agents—and a moral dimension—guaranteeing that these autonomous agents behave ethically. It will be obvious that this project raises a certain number of questions in relation to the arguments we have advanced in the preceding pages.

We have insisted that genuine substitutes cannot help but be endowed with autonomy and be capable of exercising authority. Now, authority is a moral relation, and autonomy is one of the central concepts of modern ethics.[9] The question therefore arises: does robot ethics in its current form apply to substitutes? What difference, if any, is there between an ethics for substitutes and robot ethics? Ought substitutes be taught how to tell right from wrong? And what could this really mean, teaching a robot how to tell right from wrong? Before trying to answer these questions, it will be instructive to look a little more closely at what exactly robot ethics is trying to achieve.

It may seem paradoxical, or at least curious, that the chief advocates of robot ethics tell us we are still very far from being able to create autonomous artificial agents capable of being true moral agents. Indeed, they say, we do not know whether one day we are likely to succeed or not. We do not even know if such machines are possible.[10] And so we are entitled to wonder, what is the point of the exercise? Why create an ethics for machines that do not exist and may never exist?

The reply given to this objection by robot ethicists is in two parts, which together make it fairly clear that their real purpose is not to devise ethical rules for artificial agents or to create artificial moral agents. It is in fact something quite different. The first part of the reply comes to this: we should not wait until we have been caught off guard by the sudden emergence of autonomous artificial agents to draw up such rules, when it will already be too late; it is imperative that we begin at once to prepare ourselves for an *inevitable* future. Philosophers have a part to play in the responsible development of powerful machines that very soon will be widespread in the world, by helping to figure out how to equip robots with *built-in moral rules that constrain their behavior.* The second part of the reply consists in saying that there already exist autonomous artificial agents that make decisions having potentially grave consequences, on financial markets, for example, as well as in civil aviation and military combat, and that it is of the utmost importance that these decisions be morally informed. It is essential, then, that these autonomous agents be equipped with *built-in moral rules that restrict what they are capable of doing.* The two parts of the reply are complementary. Together they indicate that the aim of the enterprise, to paraphrase Kant, is to ensure that autonomous artificial agents will "act in accordance with morality," not that they will "act morally." There is nothing overly subtle about the difference between these two ways of acting. Moreover, as we will see, the difference between them has fundamental political consequences, for it is inseparable from the question of the type of autonomy with which artificial agents are to be endowed.

What robot ethics aims to do is to frame and limit the social consequences of a process of technological development whose future course is judged to be both inevitable and predictable, and, indeed, inevitable because predictable. This is because we can already tell what these machines will turn out to be, can already determine which roles they will be asked to play and what advantages they will bring; and it is because these advantages are considered to be economically self-justifying that nothing, it is believed, apart from a catastrophe or a highly unlikely turn of political events, can stand in the way of their becoming a reality. Robot ethics seeks to curb and circumscribe the transformations entailed by a generalized use of robots *before they take place.*[11] Its purpose is not to invent new ethical rules specifically pertaining to robots, in the fashion of Isaac Asimov's three fundamental laws of robotics, but to apply to the behavior of autonomous artificial agents ethical injunctions drawn from an accepted moral system, such as utilitarianism.[12] The idea is to restrict what robots and autonomous machines can do from within, as it were, by incorporating an "ethics module" in their command system. This module contains all those ethical rules that autonomous artificial agents must consult before acting, rules whose goal is to limit their freedom of action.

This strategy reflects a fundamental ambivalence toward the autonomy of artificial agents, which we considered briefly in the Introduction. It arises from the fact that, by and large, we do not wish to create *truly* autonomous agents. It also reflects a certain confusion regarding the role of autonomy in moral philosophy. We are told, on the one hand, that there do not exist artificial agents that are sufficiently autonomous to be genuinely moral agents, and, on the other, that there exist artificial agents whose capacity for autonomous action poses moral problems and requires the application of ethical rules. These rules are intended to make artificial agents "moral," in other words, by restricting their freedom of action, which is nonetheless also supposed to be too limited for them to be genuinely moral agents! This confusion, it must be said, is by no means wholly unfounded. It is bound up with a fundamental aspect of modern theories of moral philosophy, which since the eighteenth century have in varying degrees been doctrines of autonomy and freedom. These theories imply, by definition, that a moral agent is someone who has the ability *not to act morally.*[13] This possibility is inherent in the very idea of "ought." An agent who acts as he ought to act is, by definition, someone who could act otherwise. If an agent cannot act otherwise, if he is constitutionally incapable of performing an action that is not moral, he does not act as he ought to act, but as he can.

Now, robot ethics denies robots autonomy in just this basic sense. It refuses to grant moral autonomy to autonomous artificial agents. This is why we are not dealing here so much with an ethics as a code, a set of rules prescribed for subjects who are not only deprived of rights, but deprived—by the very same rules that are prescribed, or rather imposed upon them—of all possibility of acting otherwise. They are deprived by these very rules of the ability to violate them. The application of robot ethics to military situations, which naturally lend themselves to such treatment, illustrates this self-nullifying conception of the autonomy of so-called autonomous artificial agents perhaps more clearly than any other. In replacing the ability to choose by the impossibility of behaving otherwise, it shows how far robot ethics diverges from modern philosophies,[14] which themselves, historically, are inseparable from the genesis of the nation state and of political liberalism.[15]

Autonomous Weapons and Artificial Agents

As the political scientist Armin Krishnan reminds us, "autonomous weapons" have long been with us, at least since the First World War.[16] An autonomous weapon is one that by itself is capable of inflicting damage on an adversary, that is, without human intervention. More precisely, any system that aims to eliminate the human operator, wholly or in part, and that once activated or released into the environment no longer requires human intervention to select its targets or to attack them is an autonomous weapon.[17] Consequently, any automated weapon or defense system, from antipersonnel mines to unmanned aerial vehicles (drones) and antiballistic missile batteries, constitutes an autonomous weapon. Within the class of autonomous weapons, robots designed for strictly military purposes form a particular subclass. Krishnan defines a robot as any programmable machine that is capable of sensing and manipulating its environment and that is endowed with a minimum of autonomy. In this respect, a robot is distinct from an autonomous weapon such as an antipersonnel mine. Since a mine is limited to detonating automatically when it is disturbed, it is incapable of perceiving or controlling its environment.[18] It follows from this definition of robots as a particular type of autonomous weapon that not all autonomous weapons are in fact autonomous. By virtue of what characteristic, then, can they properly be said to be autonomous?

Krishnan seeks to go beyond technological innovation in military affairs to analyze the ethical and legal implications of taking away two fundamental responsibilities from human operators and entrusting them to machines: the authority to decide whether to attack a target and whether the object aimed at is in fact a legitimate target. Autonomous weapons (among which military robots occupy an increasingly significant place) are therefore all those weapons that remove these responsibilities from human agents and put them in the "hands" of automated systems or artificial agents. These systems and agents are autonomous in the sense that we are incapable of continuously monitoring and regulating their behavior and that they alone determine whether to take action, which may assume the form either of aggression or retaliation.

For Krishnan, the primary ethical issue has to do with transferring the act of killing or wounding to devices that are not wholly subject to our control. In fact, there is nothing new about this phenomenon. It has existed since humans began making traps to capture or kill their enemies. What has changed is that such devices have become much more complex, intelligent, and, above all, active. Many modern devices are no longer limited to passively waiting until an enemy comes along and falls into a trap or steps on a mine. They are capable of taking the initiative themselves, of attacking or counterattacking.

What is more—and this is a central point in Krishnan's argument—autonomous devices today tend to take over military strategy. The coordination in real time of various weapons and combat units over an entire theater of operations generates a vast quantity of data that only machines are capable of analyzing; and once a situation has been analyzed, only machines are capable of deciding quickly enough what steps need to be taken in response. Individuals who make decisions, whether commanding officers on the ground or their superiors elsewhere, do so on the basis of automated procedures over which they have no control, and their action is often limited to choosing from a list of options dictated by a computer system, since the staggering amounts of information it can process are altogether beyond their own ability to absorb.[19] It certainly remains within their power to trust their own judgment and to reject the machine's conclusions. But these systems tend to do more than relieve officers of their duties; in the event of the failure of an operation, having made a decision personally rather than following a machine's advice becomes a rather more serious matter owing to the simple fact that powerful "autonomous" instruments have been placed at their service for a reason. Reliance on an automated system for managing military information and making decisions on the basis of this information sets in motion a paradoxical dynamic whereby the system absolves officers of all responsibility while at the same time making it irresponsible not to follow its recommendations, which now begin to look like orders.

As a consequence, Krishnan argues, the scope for human initiative and accountability becomes correspondingly reduced. We leave it to systems—autonomous systems, in the sense that they escape our control—to decide whether to kill, whom to kill, and under what circumstances. This is not to be interpreted as an argument against technological innovation or in favor of somehow limiting it. Krishnan holds that the development of autonomous weapons will continue to be extremely rapid, mainly for economic and political reasons.

The logic of the arms race in this domain is such that in the years ahead we will not be able to avoid handing over to autonomous systems a large share of our responsibilities and our decisions. Battlefield management will have become impossible otherwise.[20] This is why the ethical and legal consequences of this surrender are so important.

The Ethics of Autonomous Military Robots

A pair of recent books by Ronald Arkin, a specialist in autonomous robots, addresses the concerns raised by Krishnan.[21] Under what circumstances, Arkin asks, is it moral for a robot to kill?[22] He does not inquire into the legitimacy (which he takes for granted) of the tendency (which he considers to be inevitable) to transfer to autonomous systems the power to kill.[23] Indeed, he believes that there is a considerable advantage—we will soon see why—in confiding this responsibility to artificial agents rather than reserving it exclusively to human beings. Whereas Krishnan examines the ethical and legal consequences of transferring the decision to kill to autonomous systems, Arkin seeks to identify those situations in which it is ethical for artificial agents to kill and to ensure that in all other circumstances they will be incapable of doing so. In the second of the two volumes, he goes on to develop a technically sophisticated and rigorous protocol that will make it possible to appropriately regulate the violent behavior, homicidal and otherwise, of autonomous military robots.[24]

Arkin therefore is not concerned with the ethical conditions and consequences of permitting a robot to kill, of giving it the right and authority to exert lethal force. Rather, he is concerned with finding ways of limiting such behavior so that it will always be in accordance with certain rules—international laws of war and national (that is, American) military norms, particularly with regard to rules of engagement, which determine the circumstances in which it is permitted to open fire on an enemy—considered to have binding ethical authority.[25] Arkin takes it for granted, in other words, that there are situations in which the use of lethal force is legitimate and that there exist rules, generally recognized and accepted, that determine when, in what situations, and in what manner the use of such force is legitimate. His aim is to ensure that autonomous robots granted the power to kill on their own, which is to say without constant human supervision, will always perfectly respect these rules.

If, as Arkin maintains, autonomous artificial agents may prove to be morally superior to human beings in combat situations, this is chiefly for two reasons. First, they know neither fear, nor hatred, nor the desire for vengeance, nor the seductive charms of glory, and their judgment cannot be clouded by panic or rage.[26] They are, in other words, machines without passions or emotions. This, Arkin claims, is why a robot, unlike a human soldier, will never be carried away in the heat of combat, will never commit atrocities or make unreasonable decisions. The second reason is closely related to the first: unless someone directly manipulates and compromises its control system, it is impossible for a robot to do anything other than follow the ethical rules with which it has been programmed. An "ethical" robot therefore always obeys the orders it receives, unless they contravene the laws of war or the rules of engagement, which it respects in order of priority. It is possible, of course, that a military robot may make mistakes, whether owing to some technical malfunction or to the fact that the information available to it is incomplete. Nevertheless, assuming that its control system has been well designed and that it functions as it should, a robot is incapable of committing an ethical mistake in any of the situations contemplated by the doctrine of morally permissible lethal action that guides it or in any other situation for which it is capable of deducing an appropriate behavior.[27]

According to Arkin, the great advantage of artificial agents over human soldiers is that under all circumstances they will necessarily act in an ethical fashion, for *they are deprived of liberty and of the weaknesses that cause us to deceive ourselves and to fall into error.* But these weaknesses, and the liberty that is inseparable from them, are the very things that lead us to devise ethical systems in the first place. The superiority of "ethical" robots, according to Arkin, comes from the fact that once they have been programmed to behave ethically, they cannot behave otherwise. "The robot," he emphasizes, "need not derive the underlying moral precepts; it needs solely to apply them. Especially in the case of a battlefield robot (but also for a human soldier), we do not want the agent to derive its own beliefs regarding the moral implications of the use of lethal force, but rather to be able to apply those that have been previously derived by humanity as prescribed in the [laws of war] and [the rules of engagement]."[28]

In short, what we want are obedient and disciplined robots that do not think for themselves, which is to say robots that are autonomous only with respect to a restricted and clearly specified set of tasks—determining the enemy's position and the size of its forces, the appropriate response, and so on—and that scrupulously observe the laws of war and the rules of engagement. In that, too, robots are held to be superior to humans. Just as they are more reliable in battle, because they aim more accurately and fire more rapidly, make no careless mistakes, and are never afraid, so too they never violate the laws of war or the rules of engagement. They are ideal soldiers.

Arkin's argument is revealing, not least for exposing two distinct tendencies, complementary rather than contradictory, that may be discerned in the apparently inevitable course of events that Krishnan describes. First, there is the interest in relinquishing to autonomous artificial agents, rather than confiding to other human beings, the decision to kill or simply to initiate hostilities. Whereas Krishnan finds this deeply troubling, Arkin seems actually to welcome it. Second—and this is precisely what a willingness to surrender such authority only barely conceals—there is the desire to reserve executive decisions to a small number of human actors and to establish a perfect chain of command. One might say with only slight exaggeration that the "ethical" robots Arkin envisions are the soldiers that every general dreams of, for they make it possible to enforce discipline to the highest degree possible. But the appeal of soldiers that are unfailingly obedient and never make mistakes, just as in the case of automated information management systems, has still more to do with the increased power they are supposed to confer upon commanding officers, who are now able, or so it is thought, to exercise greater control and thereby avoid the fog of war.

It will be plain, then, that we are not dealing here either with a simple and straightforward course of historical development or with a situation where technological determinism leads inevitably to life-or-death decisions being left to systems that are blind to some of their essential aspects, or, if they are not completely blinkered, have at least, like Arkin's robots, no more than a narrow and single-minded view of what is at stake. That is one issue. There is also a fundamental political issue: relinquishing the power to choose to autonomous systems intensifies *the concentration of the political and moral capacity for decision* by confiding it to only a few human agents. These issues are not in any way opposed or contradictory; they go hand in hand. The moral dilemma posed by taking away from human beings the ability to decide whom is to be killed, and when, is inseparable from the pursuit and strengthening of a political aim that also constitutes a disciplinary strategy common to every army,[29] namely, reserving to only a few individuals the ability to make decisions and issue orders that are then strictly obeyed.

This tendency is not confined to military affairs. Everywhere it is observed to operate in civilian life, the process by which the power to decide is taken away from human agents and confided to artificial agents has the effect, in addition to the technological advantage of improving the ability to manage and rapidly process large quantities of information, of concentrating authority, so that decisions are entrusted to an ever smaller circle of individuals: computer scientists and business executives, in addition to military officers—in short, to an elite group of technical experts and unelected officials who formulate and apply the rules that constrain the behavior of artificial agents. Note, too, that because we frequently interact with these agents, which structure important aspects of our daily lives, answering our questions, making suggestions, and questioning us in their turn, robot ethics and machine ethics must be seen as "ethically" restricting not only the freedom to decide enjoyed by artificial agents, but also the freedom of human beings who come into contact with them. By giving machines no choice but to obey rules that limit their possible behaviors, these ethics indirectly—but no less surely for that— reduce our own ability to choose.

This is why Arkin's scheme cannot be regarded as a means solely of assisting military commanders. It aims also, again indirectly, at restricting their own room for maneuver. The inability to neglect or disobey rules that *ought to guide* human beings in making decisions, which is built into "autonomous" machines that are meant to be their servants, inevitably reduces strategic flexibility. Arkin's scheme seeks to make war moral, in other words, by regulating not only the combat behavior of robots, but also that of the officers who command them. Yet this moralizing strategy never appeals to reasons for acting in one way rather than another. It is intent instead on transforming the environment in which human beings and artificial agents act in such a way that it becomes impossible for them not to act in accordance with certain moral laws. The strategy is applied directly to artificial agents: their built-in command structure is such that they cannot do otherwise than respect these laws. It is applied indirectly to human beings—that is, to the "natural" agents who command them: the transformation brought about by the massive deployment of ethical autonomous military robots has the effect of narrowing the range of possible actions available to military leaders by forcing troops always to act morally, which is to say in perfect conformity with the laws of war and the rules of engagement.

Arkin is nevertheless aware that in the case of military command things are somewhat more complicated. Even supposing that the heads of the various services were jointly to approve his scheme, it is clear that theater commanders and field officers will always wish to reserve to themselves the ability to judge whether the specific circumstances of combat make it advisable, or actually necessary, to depart from strict respect for the laws of war. As a technical matter, of course, it is possible to make sure that a human operator can override the robot's command system to allow it to circumvent the moral limitations on its behavior. In the interest both of preserving this option in case of emergency and of protecting against abuses, Arkin recommends instituting a so-called dual-key system modeled on the control protocol for the launch of nuclear missiles. The aim is to permit the ethical constraints that weigh upon the behavior of robots to be removed while at the same time making it extremely difficult to do so.

This precaution clearly shows that the decision to kill is, in the last analysis, always made by human beings. It is made by those who impose ethical constraints on the behavior of autonomous systems and who are also capable of removing them when this appears to be advantageous. Ethical autonomous military robots are really nothing more than mechanical slaves that can do nothing other than obey and whose ethics module can always be countermanded "if circumstances require," which is to say whenever a human agent decides that they do. It is this expedient that reveals the weaknesses of Arkin's argument. His ethical scheme may be considered authoritarian to the extent that its fundamental, if not its sole, objective is to make agents act in accordance with certain moral rules; and yet it seeks to achieve this result without ever having to invoke the authority of any moral reason. To ensure that agents will always act in accordance with moral rules, Arkin proposes either to modify an agent's behavior directly, in the case of robots,[30] or to modify the environment in which an agent intervenes, in the case of human beings. The aim in the second case is to transform the decision environment in such a way that military commanders cannot act unethically.

Now, this dual purpose—of transferring the ability to kill to robots while simultaneously concentrating the power of decision in the hands of fewer and fewer human beings—is encouraged by a great many things. In the first place, there are technical considerations. Autonomous systems are more rapid, more precise, and capable of processing far more information than human beings. Then there are economic considerations. An unpiloted plane costs about a tenth of what an equivalent piloted model costs, and the cost of maintaining a robot combatant is disproportionately less than the cost of paying and providing support for a human soldier.[31]

Next are the political considerations. The death of soldiers always provokes a deep emotional response and often causes voters who were favorably disposed to official policy to question the wisdom of intervening militarily abroad or of otherwise deploying troops in distant lands. Politically, a robot is worth nothing. One can sacrifice as many of them as one likes!

Finally, there are purely military considerations. The arms race has a long history at this point. Now that it has been expanded to include military robots and other autonomous weapons, it is believed more firmly than ever that any country that refuses to take part is condemned to a position of weakness. Furthermore, the prospect of not having to share power holds a fatal charm for those who exercise it at the highest levels; indeed, the illusion of a perfectly smooth chain of command, in which no subordinate is free to do as he pleases, is almost irresistible. It constitutes, as we have seen, an army's organizational ideal.

Whether these considerations are also good reasons for simultaneously relinquishing the power to kill to artificial agents and autonomous weapons, and for reserving to only a few individuals the right to make ultimate decisions in matters of life and death, is an altogether different ethical and political question. Robot ethics, at least as Arkin understands it, does not even think to pose this question. As far as Arkin is concerned, the answer is clear and already known. It is a matter only of ensuring that the artificial agents to which this power to kill is handed over will always strictly respect the orders they receive and act in accordance with an "ethical" code, and nothing more. Outside the military domain, as we have seen, robot ethics likewise considers that surrendering the responsibility for deciding to artificial agents is inevitable, but without ever asking whether the concentration of power in human hands that follows from it is likewise inevitable. More precisely, the question that robot ethics never asks is whether this centralization of decision-making authority is politically desirable and whether it is morally justified. The plain fact of the matter is that no one, or almost no one,[32] seems to have noticed the political phenomenon of concentration of power that accompanies the deployment of technological innovations for military and other purposes.

With regard to the question that is at the heart of our book, it is significant that the superiority claimed on behalf of Arkin's presumptively ethical robot combatants is associated in large part with their lack of emotions. Yet the same impulses that Arkin fears will drive his machines to commit atrocities—as if his robots were mere humans!—are also what might permit them to feel compassion and to show respect for, or at least an interest in, something other than the flawless accomplishment of their deadly mission. A bit of artificial empathy would allow these robots, even if only to a very small extent, to be social actors rather than simply weapons. The use of autonomous robot soldiers—mechanical slaves, in effect, one-dimensional killing machines—reflects a wholly strategic concern with achieving crushing military superiority, without any thought being given to the importance of establishing relations of mutual trust and acceptance among foreign troops and local populations—relations that, as recent Western military adventures in Iraq and Afghanistan have shown, alone make it possible for military victory to bring about lasting peace and stable government.

Morality and Behavior Management

Arkin's military robot ethics is not really an ethics or a moral doctrine, but a technique of behavior management, which is something quite different. What Arkin seeks to do is ensure that artificial autonomous agents will be incapable of violating the rules to which they are subject, to guarantee that it will be impossible for them not to respect such rules. He resorts to a slightly different strategy, but nevertheless one of the same type, with regard to the class of human autonomous agents with which he is concerned, military decision makers. The point of introducing "ethical" robots on a large scale is to transform the military environment in such a way that it will be more and more difficult for officers either to act contrary to the laws of war or to disregard the applicable rules of engagement. Eventually, if all goes according to plan, it will become impossible for them not to "act morally."

From Arkin's point of view, robots are considered to act ethically inasmuch and so long as their behavior conforms to what is demanded of them by the rules of behavior that are incorporated in their control architecture. The *possible* consequences of a robot's actions are taken into account by its designers to determine which actions are or are not permitted in a particular circumstance by the rules it has been given. But so long as the condition of rule agreement is satisfied, the robot acts ethically no matter what the *actual* consequences of its actions may later turn out to be.[33] A military robot, in other words, never has any cause for regret, for its morality is a completely private matter. It is confined to the purely logical space in which the rules that have been given to it are brought into alignment with the actions it is permitted to perform. What subsequently takes place in the world as a result of its action is of no concern whatever, for it has nothing to do with military ethics.

All existing versions of robot ethics, so far as we can tell, exhibit this logical structure. What is wanted, the thing that defines the morality of an action, is that the robot follows the rule that is given to it. It does not matter how this result is arrived at so long as it is arrived at. An artificial agent that follows a rule—more precisely, one that acts in accordance with a rule—is considered to act morally. Now, the best, the surest way of consistently achieving this result is to make sure that the agent can in no circumstance act otherwise than the rule prescribes. Acting morally is thereby reduced to acting in accordance with a rule whose moral character has been decided in advance, externally, which is to say without the agent being consulted, without the agent being able to consent to the rule to which it is submitted and to affirm its morality. In this view, an agent acts no less morally whether it follows the rule freely, is forced to follow it, or is incapable for whatever reason of acting otherwise. All methods of regulating an agent's behavior, so long as they do not come into conflict with the rule itself, are morally desirable and equivalent.

Robot ethics therefore does not amount to an ethics, at least not in the modern sense of the term, in the sense given it by moral philosophies of autonomy. These philosophies suppose not only that an agent has the freedom to act otherwise, to act in a nonmoral fashion, but also that the choice between acting morally or not is available to it. This choice, and the ability to act otherwise, is exactly what makes it possible for an action properly to be said to be moral, rather than simply in accordance with morality. The sort of liberty that is assumed by modern moral philosophies of autonomy is, as Peter Strawson and Isaiah Berlin have demonstrated, entirely superficial, devoid of all metaphysical depth,[34] for it does not have to confront the question of free will. What is at issue here, then, is not a determinism to which robots are subject, but that human beings somehow manage to escape—which would explain why an ethics of liberty does not apply to robots. To the contrary, to be able to act morally one has at a minimum to be able to choose between alternative courses of action. But it is exactly this possibility, of being able to choose between alternatives, that robot ethics denies to artificial agents.

It is essential to Arkin's way of looking at the matter that an agent (natural or artificial), to the extent that he (or it) is autonomous, be unable to judge by himself the morality of the rule to which he is subject, since the only thing that matters is ensuring that he will always act in accordance with it. This condition is not peculiar to the ethics of military artificial agents. As the examples given by Wallach and Allen in *Moral Machines* clearly show, different versions of robot ethics, while they do not all agree on the same code of conduct, do nonetheless all try to do the same thing: to see to it, by providing the robot with rules of behavior, that its actions will always and under all circumstances conform to what these rules prescribe. For the moment, immense technical difficulties remain to be overcome before a robot will be able to determine, for example, which rule applies in a given situation. Even supposing that all such challenges can be met, artificial agents that flawlessly follow the moral rules that have been implanted in them will not behave ethically. They will simply be acting in accordance with a rule that may lead either to results that we judge to be ethically desirable or to results that we consider to be monstrous.

The reason why robots capable of true moral autonomy do not yet exist is not merely technical, however. As Arkin's disciplinary strategy—endorsed, it would appear, by all ethical thinking on the subject—makes equally clear, the reason has to do also, if not mainly, with the fact that these are just the sorts of robots we do not want. It therefore comes as no surprise that robot ethics is not really an ethics. It is not a question, as ethicists usually suppose, of teaching robots the difference between right and wrong, but, as the title of Arkin's most recent book revealingly urges, of *governing* the behavior of putatively autonomous artificial agents and of managing the consequences of their actions. This amounts to a frank acknowledgment that the problems raised today by reliance on such agents for decision-making purposes by people occupying positions of the highest authority are often to a greater extent legal and political in nature than they are ethical.

Many of the autonomous artificial agents that we fabricate today are not substitutes. Military robots are weapons exclusively dedicated to one or more specialized purposes: reconnaissance, destruction of the enemy, identification and selection of targets, and so on. The altogether honorable end that military robot ethics aims at is to ensure that such robots will not be immoral weapons, in the sense in which weapons prohibited by the international community and the law of war—poisonous and other lethal gases, for example, or bacteriological weapons, or certain antipersonnel mines—are immoral. But a weapon permitted by international law—an antitank missile, for example—does not therefore become a *moral* weapon! What we are dealing with here, at bottom, is a category error.

The various ethics that are concerned with artificial agents seek to manage their behavior as if they were individual actors. They claim that artificial agents can be made moral, and their behavior ethical, by imposing a set of rules that they cannot disregard or otherwise fail to conform to. The fact of the matter, however, is that they are not individual agents; they are moments of a complex technological

system, in the case of partial agents, or parts of a vast "social machine" (as an army may be thought of),[35] in the case of military robots.

The ethics that Arkin has devised aims at making military robots not merely cogs in a machine, but ideal cogs. The same goes for all other artificial agents. The various ethics that are concerned with them aim at making not merely doctors, traders, bank clerks, nurse's aides (or anyone else who acts in the name of these or other professions), but perfectly performing doctors, traders, bank clerks, nurse's aides, and so on. The purpose, in other words, is to create a class of workers that at all times and in all circumstances will respect the legal rules and the ethical code of their profession or office (or at least of a small part of the domain in which their duties lie). These ethics seek to produce model employees that will never be a source of worry for their superiors, because they will always have followed the rules to the letter.

Action and Autonomy

Autonomous agents act, and their actions have consequences in the world, some of them good, some of them less good. For example, the banking algorithm that decides whether to approve or reject your request for funds on the basis of statistical information regarding your habits may be said to act, at least in a minimal sense of the term. By means of a complex system, of which it constitutes only one of a great many links, it directs that a charge be accepted or disallowed, or even (in the case that you request cash from an automated banking machine) that your bank card be confiscated. Moralizing the system's behavior would consist, for example, in giving it access to additional information about your current financial situation and the consequences that a denial of funds would have for you and your family. It would evaluate these consequences by applying some rule—another algorithm—while also taking into account the chance of fraud. In

this case, morality would coincide with sound business practice, for it is certainly desirable from the bank's point of view, however important it may be to ensure the security of automated transactions, that clients not be denied ready access to their money or to a particular customer service when they are urgently in need of it.

But is there anything that may properly be said to be ethical in the behavior of a complex system of this sort? Two things immediately come to mind: protecting the interests of honest clients and guaranteeing that access to what is rightfully theirs is unimpeded. In fulfilling the dual purpose of preventing fraud and of reserving the sole use of a credit card to its authorized holder, the system plays at once the role of policeman and bank teller. If these people can be said to act morally or ethically when they correctly carry out their duties, why should the same not be said of an artificial autonomous agent when it does the same thing? The problem is this: the autonomous agent *does not do the same thing;* it achieves, or seeks to achieve, the same result.

Furthermore, the artificial autonomous agent is perfectly blind to this result. It "sees" only the rule that is supposed to guarantee the result and that it cannot help but follow. Consequently, as Arkin's military ethics makes unmistakably clear, surrendering to autonomous machines the ability to decide in our place amounts to surrendering the ability to decide to those persons who conceive and write the rules that limit the possible actions of these very machines. In certain cases, such as that of the banking automaton, this is normally unproblematic, for there is a social consensus about the desirability of a rule that is intended to protect us against fraud and about the importance of respecting it. Here the fundamental difficulty is technical in nature. We need to be sure that the relevant algorithm will be smart enough to see to it that the decisions it makes for the sake of our protection and security will not (or not very often) deprive us of legitimate access to our funds and will not (or not very often) fail to

recognize fraudulent transactions. In other cases, however, such as that of military robots, the situation is much more complicated. Here there may be disagreement about the reasonableness of the rules of engagement, for example, and about what sort of room for maneuver decision makers should be allowed to claim for themselves in case of emergency. But there is a deeper problem: the lack of any political debate regarding the rules that ought to limit murderous behavior in military robots and about whether such robots ought to be developed in the first place. Leaving it up to military robots to decide when and under what circumstances they can use lethal force really amounts to transferring to someone else—to other human beings, not to machines—the power to answer these questions.

As Arkin well understands, the autonomous artificial agent that decides for us prevents us from deciding ourselves—either because it prohibits us from deciding otherwise or because it makes doing this difficult since, being an autonomous agent, it will often already have acted. It will be necessary for us to countermand it, to go back and cancel its action. This is not always possible. It is precisely this difficulty that the precautionary measures recommended by Arkin are meant to dispel. By permitting a human operator to overrule the ethical constraints on military robots, it becomes possible to make the robot act otherwise than it was programmed to act, to license an action that had been forbidden to it. In the event that the shifting fortunes of combat make it likely that strict respect for the laws of war or the rules of engagement will prove disastrous, it is incumbent upon an officer, who now finds himself confronted with a genuine moral dilemma, to decide whether to permit the autonomous agent to act contrary to the accepted norms of warfare. Because combat situations change rapidly and in unforeseeable ways, armies must be able to react flexibly and without delay. This is why Arkin's solution probably stands little chance of being accepted. The procedure it requires is too cumbersome and takes too long to execute.

In civilian life, things are different. In the event that an automated banking machine takes the initiative of keeping your card on a Friday evening, there is not much you can do, apart perhaps from calling an emergency telephone number—only to be told to go to your bank branch on Monday morning! Faced with an unobliging bank officer, you can produce a whole array of proofs of identity and financial condition, you can arrange to have a copy of a crucial document sent at once to the bank, or you can demand to see the bank manager, whom you know personally. The autonomous artificial agent, for its part, is far less responsive to pleading of this sort than even the most obtuse or unsympathetic functionary. Indeed, one of the reasons for deploying autonomous agents is to make all such excuses impossible, in part because bank managers, when they deal with clients whom they know too well, tend to make errors of judgment or show a preference for private interests that do not perfectly coincide with those of the bank.

The fact that artificial agents who decide for us constrain our own ability to act, by preventing us from choosing otherwise than they choose, goes a long way toward explaining why their use is increasingly favored today. It has the further consequence that the most important questions raised by their use are not so much ethical as political. The first of these political questions is this: who decides, and by what right, which rules will, by determining the behavior of artificial agents and restricting their choices, determine our behavior and restrict our choices? The second question concerns what the anthropologist and political thinker David Graeber calls the "utopia of rules."[36] The multiplication of artificial agents entails an increase in the number of the rules and regulations that circumscribe and control our daily lives. These rules bear upon what is permitted and what is not permitted over a whole range of quite ordinary activities that used to be much less strictly regimented than they are today: buying a car, for example, or booking an airline ticket, or making a hotel reservation, or organizing a school trip with students. To be sure, artificial agents are not solely, or even primarily, responsible for the increased bureaucratization of our existence; it is bureaucratization itself that encourages our reliance on artificial agents. But this reliance, in its turn, can only have the effect of reinforcing the tendency to bureaucratization and of extending it to every aspect of life. To see this phenomenon as an essentially ethical problem, as robot ethics does, avoids its fundamentally political dimension while at the same time making virtues of bureaucratic rules.

It is neither necessary nor helpful to regard the limits imposed by autonomous agents on our ability to act as the result of a diabolical conspiracy or of technological determinism. The growing use of autonomous agents holds up a mirror to social issues and the choices we make in response to them. In the workplace, it is a consequence of trying to do one's job as well as possible, of making sure that letters sent out by one's company all have the same format, looking for ways to increase sales, assuring customers that the most popular products will always be available, protecting against theft and fraud, and so on. It is a consequence, too, of the fact that companies and governments have a particular interest in every one of their employees doing the best job possible. The bureaucratic temptation is to try to optimize job performance by requiring that everyone do the same thing in the same manner. Greater uniformity is a necessary condition of increased reliance on autonomous agents, and the inevitable consequence of this increased reliance is greater uniformity.

On the Intelligence and Autonomy of Partial Agents

One of the outstanding characteristics of many intelligent artificial agents, which sets them apart from substitutes, as we saw earlier, is that they are not individuals. Very often, in fact, they are not even physical objects. An artificial agent of this sort is generally defined as

a computational (information-processing) entity inhabiting a complex environment that it perceives, and within which it acts, in an autonomous fashion.[37] It amounts, in other words, to a few lines of
code, a computer program that may exist in any number of copies or
"instances." Moreover, it is not always possible to know which instance of a particular banking algorithm processed your request: in a
distributed system ("cloud computing"), there is usually no way to
tell where the agent acted—where the calculation was made that led
to the decision to approve or reject your request. The artificial agent
in this case is not an individual, nor is it a three-dimensional object; in
a sense, it is not even a thing. It is not a physically detectable and identifiable object.

We propose calling such integrated agents *analytic agents.* They are
analytic in the sense that they represent moments in the mathematical
analysis of a complex sequence of actions that, taken together, constitute the subprogram to which a decision is attributed. Erase this bit
of code, in other words, this part of the program, and the system will
be incapable of doing what it did before—for example, processing
your banking data. This bit of code is therefore indeed an agent, the
thing that *performs* the action in question. And yet it is clear that,
from a purely physical point of view, it is an agent that does not act.
Taken by itself, in isolation, this bit of code cannot meaningfully be
said to do anything at all. It can only be said to do something because
it is part of an integrated system that includes other intelligent agents
of the same type—the Internet, data banks, protocols governing
banking transactions, and so on—as well as many physical networks,
natural and artificial alike, among them automated banking machines
and the armored-truck personnel who stock them. These systems, of
which analytic agents form an essential element, are therefore hybrid
systems in which a wide range of agents and networks operate in
concert. It is only when an analytic agent has been built into a system
of this type that it is capable of acting. For the system furnishes the

environment in which the analytic agent "lives," the environment that it "perceives" and within which it can act in an "autonomous" fashion. It is the system as a whole, rather than the analytic agent by itself, that interacts with us through the coordinated behavior of its various elements. As a consequence, while it certainly makes sense to say that such an agent is intelligent, to say that it is autonomous is, at the very least, a misuse of language.

The central question that arises in connection with the systems in which these analytic agents function is therefore this: who is doing the acting? This question may be understood in at least two senses. First, in a legal sense: who is responsible for a particular action—for deciding whether to buy or sell, to open fire, to authorize a communication? For the time being, at least in the commercial sphere, the answer generally is the company or other entity ("the operator") that has a license to use the computer program containing the intelligent artificial agent in question. From a legal standpoint, the program is merely a tool. The operator is generally held to be responsible for any damages that this tool may cause to third parties, even in the event of the tool's malfunctioning. Only if a third party displays culpable negligence may the operator be exempted from liability. Evidently, this does not exclude the possibility that the operator may turn around and sue the manufacturer or the creator of the computer program, accusing one or both of selling him a defective product. It is interesting that the concept of legal responsibility at issue here does not require anyone to have actually done anything at all. If you are the owner of a factory and an accident causes injuries to third parties, you may be legally responsible even if you did not commit the act—you did not blow up your factory—and even if, strictly speaking, no one committed the act causing the event (in this case, an accidental explosion) whose consequences you are blamed for.[38]

The question of who is acting here may also be understood in a practical sense: who performed the action? Who sold your shares, for example, or accepted your bid to buy shares online through a broker website? Who decided to buy your shares or to engage in electronic financial transactions? The answer seems to be no one in particular, or else the "system" as a whole, for at the subsystem level it is not always possible to associate an identifiable person or entity with the analytic agent in question.[39] This is why one must accept that, in a certain sense, there is neither actor nor action here. An analytic agent is not an actor. It is an element of a complex system, a component, designed and implemented in the system so that events will occur in the world that correspond to, resemble, or at least have something recognizably in common with other events that in the past were the consequence of human actions or that under other circumstances are the consequence of human actions still today.

Now, these authorless events that intervene in our lives in something like the way that ordinary actions do are nonetheless liable themselves to have major consequences. Beyond whatever intrinsic significance they may have, the decisions made by artificial agents that act on our behalf and decide in our place reflect, as we have seen, a tendency for decision making to be reserved to an increasingly restricted number of human actors. For technical reasons bound up with the distinctive nature of analytic (or partial) agents and what they can "do," but also for economic reasons, the concentration of the power to decide in fewer and fewer hands heralds the advent of ever larger, more complex, and more interdependent systems, particularly in connection with the management of financial markets, transportation networks, information, and manufacturing and distribution.

As the sociologist Charles Perrow has shown, this state of affairs unavoidably carries with it an element of hazardous risk. Beyond a certain threshold of complexity, accidents will inevitably occur, for we cannot entirely predict the behavior of a complex system, still less regulate it in every detail.[40] This inability to anticipate how the system as a whole will behave is precisely one of the reasons why we are tempted to place our faith in intelligent artificial agents. The more concentrated power becomes in a system, and the more tightly coupled the many subsystems that it integrates, the more disastrous and long-lasting the consequences of an accident are likely to be. One hardly needs to imagine that systems incorporating many intelligent artificial agents will one day become conscious to understand how easily they can become autonomous—that is, how apt they already are to escape our control. No so-called singularity, no metaphysical catastrophe is required to explain what Perrow calls "normal accidents."[41]

The apocalyptic fear of a revolt by machines, which we discussed at the beginning of this book, is a sign of the ambivalence we feel toward these invisible and omnipresent analytic agents. We use them to decide for us and to serve us. Our fear betrays—and translates into another language, that of a technology that now threatens to become its own master—an unintended consequence that we have yet to come to terms with: in using machines to decide for us, we have unwittingly transferred the ability to decide to other people, fewer and fewer of them, who in turn believe themselves to be correspondingly more powerful, though their power is increasingly constrained by the very rules they have given to these machines. This fear also testifies to the persistence of our instinctive dualism—the metaphysical fantasy of a naked, disembodied, self-replicating mind capable of taking refuge in the most different bodies imaginable.

Substitutes Revisited

The substitutes we considered earlier are technological objects of a very different type than analytic (or partial) agents. They are three-dimensional physical objects that are spatially recognizable as independent individuals, even in the case of semiautonomous robots. They are actors to which it is always possible to attribute, if not full authorship, at least responsibility for having performed, committed, or carried out an action. Moreover, substitutes are not exclusively dedicated to a particular end. Even when they fulfill a precise function, a therapeutic function, for example, the desired objective always requires that a specific and overtly social relationship be established between the robot and its human partner. Consequently, the ethical and political problems posed by these newcomers to our world are of another kind as well. They cannot be resolved by simply applying rules of the sort that order a company's dealings with its customers, for example, or soldiers' relations with their officers. Designing substitutes forces us to confront fundamental ethical questions raised by interaction between human beings and social robots. We believe that by honestly facing up to these questions it may also be possible to think about ethics itself in a new and different way, perhaps even to transform the nature of moral inquiry.

In Chapters 3 and 4, we tried to show that creating artificial agents capable of taking part in a dynamic of affective coordination means that they must be able to respond to the displays of emotion, the preferences and inclinations, the moods, the needs, and the desires of the persons with whom they interact. Additionally, these robots must be able to adapt to the evolving demands of the specific task they were designed to perform (as nurse's aides, companions, and so on). This evolution depends not only on the history of the robot's relationship with its human partner, but also on the changing character of its partner's needs. In the case of robots, of course, these capacities will be unevenly distributed, just as they are among human beings; and just as some people are manifestly better at certain things or socially much more adept than others, not all robots will be equally gifted or skilled, and not all will enjoy the same success with every human being they encounter.

The challenge, then, is to create robotic subjects that can sustain a flourishing human-robot dynamic because they have learned to

welcome the personal requests, respond to the characteristic expressions of emotion, and satisfy the individual needs of human
beings. If they can do these things, we will one day recognize robots
as genuinely social partners whose presence is not only convincing,
but also pleasing and positive. This dual requirement occupies a central place in research guided by the methodological principle of an
affective loop, particularly in relation to the construction of "personal robots."[42] Once removed from its box, so to speak, and placed
in a new social environment, the robot must be capable of getting
along with all those who share this environment with it. It must be
able, in the course of repeated interactions with the same person, to
develop a "personality" of its own that makes it seem like an old
acquaintance or even a close friend or relative. Substitutes must be
able, in other words, to know how to adapt to the specific circumstances of a relationship and to develop a style of interacting that is
well suited to it. This is one reason why one must suppose that their
success will vary from case to case, for the outcome of a relationship
never depends only on one of the partners.

If it is decided to equip substitutes with an ethics module con-
straining their capacity for action, they will be able to be authentic
social partners only if they are allowed a certain amount of leeway,
which is to say the spontaneity that is indispensable to companionship. For without this element of freedom, they will be unable to
develop different forms of affective coordination with different human
agents. From this it follows that an ethics module of this sort is not
reducible to a set of moral rules that cannot be violated under any
circumstances. The actions that emerge in the course of coordination
need, of course, to be kept within certain acceptable limits, but the
robot's capacity for inventiveness must nonetheless be real if it is to
be able, in the best case, to adjust its behavior to the individual
needs of a range of partners—men, women, children, the elderly,
or other artificial agents. This means, among other things, being

able to interact with "him" or with "her," rather than simply with this one or that one.

Although neither sociability nor ethical behavior can be adequately modeled by a set of predetermined rules, the mechanism of affective coordination, which underlies the concept of artificial empathy we have developed in the preceding pages, does imply that at least a rudimentary form of *ethos* must somehow be incorporated in the ability of artificial agents to respond appropriately to the emotions expressed by their human partners.[43] If the robot's capacity for empathic response is to be correctly conceived and implemented, if the social presence of substitutes is to have real meaning for their partners, it will be necessary not only that these robots manage to regard the feelings and needs of their partners as part of a broader interest in behaving morally, but also that this interest be reciprocated. While strict reciprocity may be too much to hope for at this stage, an ethical sensibility will have to make itself felt on both sides if we want our robots to be something other than mechanical slaves. A relationship of reciprocal coordination between two agents cannot properly be said to be ethical if the moral dimension of the relationship does not concern and influence both parties. Our dealings with social robots will therefore have to lead to the appearance among us of an interest in the welfare of our artificial companions, like the interest we take in the welfare of our animal companions. Coevolution between humans and truly social robots necessarily involves some amount of invention and innovation—what might be called ethical discovery. Rather than forcing us to act morally, as current robot ethics would like to do, interactions with the class of artificial agents we call substitutes will enlarge the range of situations in which it is possible to act morally and, of course, in which it is also possible to act immorally.

In the case of genuinely social robots, of empathic artificial agents in the true sense, the ethical question cannot be considered to have been settled in advance. It is not simply a matter of imposing on artificial agents rules of behavior that we have defined a priori as moral. To be sure, such rules are often indispensable, for they make it possible to avoid various kinds of danger, some of which are more foreseeable than others. They are nonetheless not sufficient to accommodate the open-ended and creative dimension of these robots' behavior. Now, even when those aspects of their behavior that cannot be preprogrammed are limited, unforeseen ethical problems will inevitably arise as a consequence of the opportunities presented by human-robot interaction. It will be all the more urgent and difficult to try to anticipate them as the partners of these robots, young and old alike, are typically disabled or disadvantaged in one way or another.

The challenges that empathic robotic agents, even those of the present generation, pose for philosophical reflection and scientific investigation provide a welcome chance to apply the synthetic methodology we described in Chapter 3.[44] Substitutes have already begun to bring about a novel type of coordinated social interaction that makes it possible to study the codetermination of ethical choices, values, and behaviors that takes place in the course of social and affective encounters. With the help of these new experimental instruments, we are now in a position to conduct empirical research on the ethical perplexities that human interaction with artificial social agents will increasingly present in the years ahead. The type of social interaction that social robots are already beginning to establish allows us to study the codetermination of values and ethical behavior as it takes place in ongoing social relations. Social robots are scientific instruments that permit us to begin an empirical and experimental inquiry into the difficulties and ethical opportunities present in our interactions with artificial agents. Alongside cognitive science, ethology, psychology, and synthetic philosophy of mind, we therefore urge that a new discipline be recognized: "synthetic ethics".[45]

Synthetic ethics takes a very different approach from the one adopted by current experimental studies on the ethical dimension of human-robot interaction.[46] These studies are concerned with the ethical challenges posed by the design, production, and growing use of robots. They aim primarily at deciding who should be consulted in this connection. In the case of medical robots, for example, they include doctors and nurses, patients, their friends and family, and so on. These studies also seek to determine appropriate ethical principles. Here they include autonomy, beneficence (action for the patient's benefit), nonmaleficence (avoidance of harm), and justice.[47] Looking to the long term, it is generally agreed that it will be important not only to show due regard for best practices, but also to cultivate social acceptance.

Synthetic ethics, while recognizing the value of these studies, looks to do something else altogether. Our view is that it is essential not to make ethical thinking about the therapeutic applications of social robotics in relation to standard forms of treatment the prerogative solely of human actors, particularly in connection with persons who have special needs. The scientific appeal of synthetic ethics is that it focuses on the mutual influence that machine and user, human agent and robotic agent, exert on their respective ethical behaviors. What needs to be studied experimentally is the possibility that these behaviors may change in the course of interaction, as well as the manner in which emergent variants have a feedback effect on the limits imposed on the robot's behavior by a previously determined set of ethical options. At bottom, what matters is the *coevolution* of the behaviors of human beings and artificial social agents. If, as we have every reason to believe, novel and unforeseeable forms of ethical (or unethical) coordination are liable to spontaneously emerge, develop, and become stabilized with the spread of interaction between humans and robots, studying these transformations will be of fundamental importance if we hope one day to understand why we think and act as we do. The goal is to anticipate and to channel—to the extent that it is possible to do so—the ethical impact of robots on our social ecology.

Synthetic ethics lays emphasis on monitoring and tentatively assessing current developments, rather than judging them on the basis of existing theories that are assumed to be adequate on the whole and that have only to be adapted in minor ways to new circumstances, when in fact it is exactly these new circumstances that call their adequacy into question. Synthetic ethics takes seriously the idea that ethical innovation is both possible and desirable. In contrast to the dark visions of a dystopian future propagated by science fiction, synthetic ethics holds out the prospect that the introduction of social robots may be, not the beginning of the end, but the way forward to a deeper and truer understanding of our nature as human beings. There may be forms of ethical reflection and of inquiries into who we are that only the further development of artificial social partners—of substitutes—will make possible.

Notes

INTRODUCTION

1 See Jason Falconer, "Panasonic's Robotic Bed / Wheelchair First to Earn Global Safety Certification," *New Atlas*, April 15, 2014, http://www.gizmag.com/panasonic-reysone-robot-bed-wheelchair-iso13482/31656/.

2 The acronym of Čapek's play stands for *Rossumovi Univerzální Roboti*, with the English translation of this phrase appearing as part of the title in the original Czech edition; among recent versions in English, see *R.U.R. (Rossum's Universal Robots)*, trans. Claudia Novack (New York: Penguin, 2004). Karel Čapek's brother, Josef, is said to have suggested the Czech word *robota* as a name for the artificial creatures of his play. Derived from an older term meaning "slave," it denotes forced or compulsory labor of the kind associated with the feudal obligation of corvée.

3 Except, of course, the robotized vacuum cleaner known as Roomba.

4 See, for example, Wendell Wallach and Colin Allen, *Moral Machines: Teaching Robots Right from Wrong* (New York: Oxford University Press, 2008).

5 In Algis Budrys, *The Unexpected Dimension* (New York: Ballantine, 1960), 92–109. In Budrys's story, the robot's revolt has to do not only with the fact of its independence, but also with the efforts of the military authorities to deprive it of the autonomy with which its creator endowed it. At the heart of the story is the conflict between a scientist who wishes to create a truly autonomous robot and an officer who is determined to stop him. We shall return to this debate in Chapter 5, which deals with military robots in particular and the ethics of robots in general.

6 Bill Joy, "Why the Future Doesn't Need Us," *Wired*, April 2000, http://www.wired.com/2000/04/joy 2/, reprinted in Deborah G. Johnson and Jameson M. Wetmore, eds., *Technology and Society: Building Our Sociotechnical Future* (Cambridge, Mass.: MIT Press, 2009), 69–91.

7 See David Dowe, "Is Stephen Hawking Right? Could AI Lead to the End of Humankind?," *IFLScience*, December 12, 2014, http://www.iflscience.com/technology/stephen-hawking-right-could-ai-lead-end-humankind.

8 See Samuel Gibbs, "Elon Musk: Artificial Intelligence Is Our Biggest Existential Threat," *Guardian*, October 27, 2014.

9 See Kevin Rawlinson, "Microsoft's Bill Gates Insists AI Is a Threat," *BBC News Technology*, January 29, 2015, http://www.bbc.com/news/31047780.

10   Should these companions be male, female, or both? The fact is that until now social robotics has avoided, or perhaps simply failed to pay attention to, gender. No serious thought has been given to how a robot that is identified as feminine might differ from one that is identified as masculine, nor has any attempt been made to implement and take advantage of this difference in order to better understand the nature of human sociability. Among robots having a human appearance, certainly there are some, such as Geminoid F and Saya, that have feminine characteristics, but these are treated for the most part in an aesthetic manner and reflect what are imagined to be the expectations of potential users (both of these robots typically function as hostesses in public settings). Some will find this state of affairs regrettable; others may argue that the absence of gender—neutrality, as it might be called—constitutes an essential aspect of artificial sociability. However this may be, we have elected to use the masculine form exclusively in referring to gender-neutral artificial agents. [Because all the artificial agents discussed in this book are asexual machines, it seems more natural in English to refer to them individually by the neuter pronoun "it." Where a specifically masculine or feminine personality is ascribed to an agent, this quality has been made clear by the relevant context.—Trans.]

11   See Hannah Arendt, *The Human Condition* (Chicago: University of Chicago Press, 1958), 7–8.

12   See Rolf Pfeifer and Alexandre Pitti, *La révolution de l'intelligence du corps* (Paris: Manuella, 2012).

13   See L. Damiano, P. Dumouchel, and H. Lehmann, "Towards human-robot affective co-evolution: Overcoming oppositions in constructing emotions and empathy," *International Journal of Social Robotics* 7, no. 1 (2015): 7–18.

14   See, for example, A. Sharkey and N. Sharkey, "Granny and the robots: Ethical issues in robot care for the elderly," *Ethics and Information Technology* 14, no. 1 (2012): 27–40, and "The crying shame of robot nannies: An ethical appraisal," *Interaction Studies* 11, no. 2 (2010): 161–190; see also R. Sparrow and L. Sparrow, "In the hands of machines? The future of aged care," *Minds and Machines* 16, no. 2 (2006): 141–161, and Sherry Turkle, *Alone Together: Why We Expect More from Technology and Less from Each Other* (New York: Basic Books, 2011).

15   See Paul Dumouchel, *Émotions: Essai sur le corps et le social* (Paris: Les Empêcheurs de Penser en Rond, 1999).

16   See T. P. O'Connor, *Animals as Neighbors: The Past and Present of Commensal Species* (East Lansing: Michigan State University Press, 2013).

1. THE SUBSTITUTE

1   While military robotics is not generally considered to be part of social robotics, some military artificial agents are manifestly social robots, and many others undeniably have a social dimension.

2   See M. Mori, "Bukimi no tani" [The Uncanny Valley], *Energy* 7, no. 4 (1970): 33–35. An English version by Karl F. MacDorman and Takashi Minato was first published in 2005, having circulated informally for a number of years; a subsequent translation by MacDorman and Norri Kageki, authorized and reviewed by Mori himself, is now available via http://spectrum.ieee.org/automaton/robotics /humanoids/the-uncanny-valley. [The title of Mori's article is meant to suggest not only a sense of strangeness, bordering on eeriness, but also the unease that accompanies it; the term "uncanny" itself recalls Freud's 1919 essay "Das Unheimliche."—Trans.] Mori's conjecture has recently been given a formal analysis by R. K. Moore, "A Bayesian explanation of the 'uncanny valley' effect and related psychological phenomena," *Nature Science Reports* 2, rep. no. 864 (2012).

3   This is the strategy adopted by Hiroshi Ishiguro at the University of Osaka, who constructs androids that appear to be copies of particular individuals, his own double among others. See K. F. MacDorman and H. Ishiguro, "The uncanny advantage of using androids in cognitive and social science research," *Interactive Studies* 7, no. 3 (2006): 297–337; also Emmanuel Grimaud and Zaven Paré, *Le jour où les robots mangeront des pommes: Conversations avec un Geminoïd* (Paris: Petra, 2011).

4   See René Girard, *Violence and the Sacred,* trans. Patrick Gregory (Baltimore: Johns Hopkins University Press, 1977); for a convergent (though wholly distinct) hypothesis on this point, see Niklas Luhmann, *A Sociological Theory of Law,* trans. Elizabeth King and Martin Albrow (London: Routledge and Kegan Paul, 1985).

5   It is interesting to note that therapists who use humanoid robots in treating autistic children do not want these machines to resemble actual human beings too closely, since the advantage they seek to exploit is bound up with the fact, or so they maintain, that it is much less disturbing for autistic children to interact with a robot than with a person. See J.-J. Cabibihan, H. Javed, M. Ang, Jr., and S. M. Aljunied, "Why robots? A survey on the roles and benefits of social robots for the therapy of children with autism," *International Journal of Social Robotics* 5, no. 4 (2013): 593–618.

6   Philippe Pinel, *A Treatise on Insanity,* trans. D. D. Davis (Sheffield, U.K.: W. Todd, 1806), 116–117; for the full discussion of cranial defects, see 110–133.

7   Pinel, *Treatise on Insanity,* 116 (note c).

8   See, for example, John Cohen, *Human Robots in Myth and Science* (London: Allen and Unwin, 1966), 15–26; also C. Malamoud, "Machines, magie, miracles," *Gradhiva* 15, no. 1 (2012): 144–161.

9   See Cohen, *Human Robots in Myth and Science*, 21.

10 Günther Anders, *L'obsolescence de l'homme: Sur l'âme à l'époque de la deuxième révolution industrielle*, trans. Christophe David (Paris: Ivrea, 2001), 37; italics in the original. [Anders's work has yet to appear in English. The French version cited here translates *Die Antiquiertheit des Menschen: Über die Seele im Zeitalter der zweiten industriellen Revolution* (Munich: C. H. Beck, 1956); a second and final volume under the same title followed in 1980 and was brought out in French by Éditions Ivrea in 2011.—Trans.]

11 That the economy as a whole may or may not, as a consequence, create as many or more jobs than were lost to robots is another question, of course, wholly separate from the purely technological function of industrial (or social) robots.

12 See http://www.parorobots.com/.

13 Paro manages this remarkable technical feat in a minimalist fashion—or, if you like, by cheating, since to the extent that it is not an actor, it is only partially and imperfectly a social being. On this point, see Paul Dumouchel and Luisa Damiano, "Artificial empathy, imitation, and mimesis," *Ars Vivendi* 1, no. 1 (2001): 18–31; also available via http://www.ritsumei-arsvi.org.

14 On this aspect of coordination, which is both prior to any cooperation or competition among agents and the condition of either one coming into existence, see Paul Dumouchel, *Émotions: Essai sur le corps et le social* (Paris: Les Empêcheurs de Penser en Rond, 1999), particularly chapters 2 and 3.

15 See Bruno Latour, *Petites leçons de sociologie des sciences* (Paris: La Découverte, 2006). For Latour, social reality does not exist apart from assemblages of humans and technological objects (or, more precisely, humans and nonhumans). In most of the examples Latour gives—seat belts, door keys, automated door closers—it is a matter of replacing a human being with a nonhuman instrument or mechanism to reduce the uncertainty associated with human action.

16 See Langdon Winter, "Do Artifacts Have Politics?," in Deborah G. Johnson and Jameson M. Wetmore, eds., *Technology and Society: Building Our Sociotechnical Future* (Cambridge, Mass.: MIT Press, 2009), 209–226.

17 This tendency can be recognized in a great many technological objects, even very simple ones. Thus a good tool remains in the background, hidden, as it were, by the hand of the artisan who skillfully manipulates it. The same is true of many weapons, particularly ones used in close combat, which appear to the soldiers who wield them as an extension of their bodies rather than as discrete ob-

jects. Today, the neurological bases of this impression are well understood; see, for example, A. Iriki, M. Tanaka, and Y. Iwamura, "Coding of modified body schema during tool use by macaque postcentral neurones," *Neuroreport* 7 (1996): 2325–2330; and Giacomo Rizzolatti and Corrado Sinigaglia, *Mirrors in the Brain: How Our Minds Share Actions and Emotions*, trans. Frances Anderson (Oxford: Oxford University Press, 2008). The same tendency is encountered in expendable products such as food processors, microwave ovens, automatic exterior lighting systems, gate openers, and lawn mowers. Unlike traditional tools that require more or less constant attention on the part of their users, technology designed for mass consumption aims as far as possible to be able to function by itself while requiring a minimum of maintenance. Not only must expendable products of this kind not be cumbersome, they should operate as invisibly and as quietly as possible. What might be called the disappearance of a product, its silence (in the sense that one speaks of health as the "silence of the organs"), is typically considered in the case of technological objects as being equivalent to efficiency.

18   And what it allows to be done to us in the way of surveillance—following our movements, for example, or verifying our identity at any moment.

19   A computer is also a three-dimensional object, but the virtual agents projected on the screen are, like shadows, two-dimensional agents. On the subject of the two-dimensional character of shade and shadow, see Roberto Casati, *The Shadow Club: The Greatest Mystery in the Universe—Shadows—and the Thinkers Who Unlocked Their Secrets*, trans. Abigail Asher (New York: Knopf, 2003).

20   See Nikolas Rose's discussion of "the ethical problems that arise in the relation between experts and their clients when trust in numbers replaces other forms of trust—that is to say, when decisions as to action seem to arise from judgments 'black boxed' within an 'objective' calculating device—whose authors are not available for debate and contestation"; *The Politics of Life Itself: Biomedicine, Power, and Subjectivity in the Twenty-First Century* (Princeton, N.J.: Princeton University Press, 2007), 75.

21   Thus traffic lights and signs posting speed limits or parking restrictions advertise an authority relation, whereas the speed bumps ("sleeping policemen," as they are sometimes called) that force cars to slow down institute a different kind of relation—an important distinction that Latour's programs of action and antiprograms fail to notice; see *Petites leçons de sociologie des sciences*, 15–75, and the note immediately following.

22   Bruno Latour, "Where Are the Missing Masses? The Sociology of a Few Mundane Artifacts," in Johnson and Wetmore, eds., *Technology and Society*, 174; emphasis in the original. A different version of this article had appeared earlier as

the first chapter of Latour, *Petites leçons de sociologie des sciences*, under the title "Petite sociologie des objets de la vie quotidienne."

23 This is manifestly the calculation made by a hotel that decides to replace a bellhop with an electronic system for automatically opening and closing the doors of guests' rooms.

24 See David McFarland, *Guilty Robots, Happy Dogs: The Question of Alien Minds* (Oxford: Oxford University Press, 2008).

25 McFarland, *Guilty Robots, Happy Dogs*, 171.

26 See Humberto R. Maturana and Francisco J. Varela, *De máquinas y seres vivos: Autopoiesis, la organización de lo vivo* (Santiago: Editorial Universitaria, 1973; 6th ed., 2004). The idea of autonomy introduced by Maturana and Varela relies on the concept of "autopoiesis," which refers to the activities of self-production and self-regulation engaged in by biological organisms, in addition to distinguishing themselves from their environment. In this connection, see Luisa Damiano, "Co-emergencies in life and science: A double proposal for biological emergentism," *Synthese* 185, no. 2 (2012): 273–294.

27 To the extent, of course, that one is prepared to grant that a lawn mower can have motivations and mental states in the first place, which may be doubted.

28 Plainly, mental states are under our control only in an extremely narrow and limited sense. We usually do not know why we are thinking about what we are thinking about. When it does seem clear to us that we are in control, when, for example, we wonder what we should make for dinner, the available options tend to be known and few in number. We are not free to modify them at will without changing the state of the world—which suggests that, generally speaking, our mental states are not under our mental control, but under the control of the world, that is, of our social, material, cultural, and intellectual environment. When, to the contrary, there are a great many options available to us, when we are thinking creatively, we do not know why we think of something that, strictly speaking, we are the first ones to discover.

29 As Kant rightly saw, it is solely on the basis of such regularities that autonomy, understood in a normative sense, can subsequently be constructed. The categorical imperative requires that the maxim of my action can be erected into a universal law. The moral actor, in this view, gives himself his own law, but he does not give himself the maxims that he subsequently erects into a universal law. Either the actor finds such maxims ready-made in his culture or society, or he derives them from an analysis of his own action, but in both cases they are part of an environment that is given to him.

30 The disjunction thus constitutes an exception, but not just any exception. It must be possible to make sense of the event to which it corresponds in relation to a rule, in the absence of which it cannot be seen as an exception.

31   In the case of a semiautonomous robot, the situation is complicated by the fact that another question now arises: who is acting, the robot or the operator? The answer will vary depending on the circumstances. Whatever it may turn out to be, however, identifying the actor is not to be confused with deciding whether the actor is an autonomous social agent.

32   The idea that autonomy implies some degree of dependence on an environment, rather than absolute independence and liberty, is inspired by research on the self-organization of living systems that seeks to account for the autonomy of human and other forms of cognition. See, for example, Maturana and Varela, *De máquinas y seres vivos*; Francisco J. Varela, *Principles of Biological Autonomy* (New York: North Holland, 1979); and Paul Dumouchel and Jean-Pierre Dupuy, eds., *L'auto-organisation: De la physique au politique* (Paris: Seuil, 1983). For a survey of more recent developments and, in particular, the application of the concept of autonomy in robotics, see David Vernon, *Artificial Cognitive Systems: A Primer* (Cambridge, Mass.: MIT Press, 2014).

33   Cohen, *Human Robots in Myth and Science*, 137.

34   On the lack of continuity between our behavior and reaction toward virtual agents, on the one hand, and three-dimensional objects on the other, see Paul Dumouchel, "Mirrors of Nature: Artificial Agents in Real Life and Virtual Worlds," in Scott Cowdell, Chris Fleming, and Joel Hodge, eds., *Mimesis, Movies, and Media*, vol. 3 of *Violence, Desire, and the Sacred* (New York: Bloomsbury Academic, 2015), 51–60.

35   See Francisco J. Varela, Evan Thompson, and Eleanor Rosch, *The Embodied Mind: Cognitive Science and Human Experience* (Cambridge, Mass.: MIT Press, 1991).

36   See Shaun Gallagher, "Interpretations of Embodied Cognition," in Wolfgang Tschacher and Claudia Bergomi, eds., *The Implications of Embodiment: Cognition and Communication* (Exeter, U.K.: Imprint Academic, 2011), 59–70.

37   The phrase is credited to Andy Clark and David Chalmers, "The extended mind," *Analysis* 58, no. 1 (1998): 10–23; reprinted in Richard Menary, ed., *The Extended Mind* (Cambridge, Mass.: MIT Press, 2010), 27–42.

38   An excellent and very ambitious example of this approach is Eric B. Baum, *What Is Thought?* (Cambridge, Mass.: MIT Press, 2004). A popular version of it may be found in the 2014 film *Transcendence*.

39   On the expansive interpretation, see, for example, Lambros Malafouris, *How Things Shape the Mind: A Theory of Material Engagement* (Cambridge, Mass.: MIT Press, 2013).

40   For an analysis of the philosophical implications of this aspect of mind, see Frédéric Worms, *Penser à quelqu'un* (Paris: Flammarion, 2014).

2. ANIMALS, MACHINES, CYBORGS, AND THE TAXI

1   See, for example, Owen Holland and David McFarland, eds., *Artificial Ethology* (Oxford: Oxford University Press, 2001).

2   It sometimes happens, rather frequently in fact, that an industrial or military use is subsequently found for these robots or for very slightly different models of which they are prototypes.

3   Among other works on the links between philosophy of mind and cognitive ethology, see Daniel Dennett, *Kinds of Minds: Toward an Understanding of Consciousness* (New York: Basic Books, 1996); Colin Allen and Marc Bekkof, *Species of Mind: The Philosophy and Biology of Cognitive Ethology* (Cambridge, Mass.: MIT Press, 1997); Joëlle Proust, *Comment l'esprit vient aux bêtes: Essai sur la représentation* (Paris: Gallimard, 1997); and Robert W. Lurz, *The Philosophy of Animal Minds* (Cambridge: Cambridge University Press, 2009).

4   See B. Webb, "Can robots make good models of biological behavior?," *Behavioral and Brain Sciences* 24, no. 6 (2001): 1033–1050; see also Luisa Damiano, Antoine Hiolle, and Lola Cañamero, "Grounding Synthetic Knowledge: An Epistemological Framework and Criteria of Relevance for the Synthetic Exploration of Life, Affect, and Social Cognition," in Tom Lenaerts, Mario Giacobini, Hugues Bersini, Paul Bourgine, Marco Dorigo, and René Doursat, eds., *Advances in Artificial Life, ECAL 2011* (Cambridge, Mass.: MIT Press, 2011), 200–207.

5   See Frank Grasso, "How Robotic Lobsters Locate Odour Sources in Turbulent Water," in Holland and McFarland, eds., *Artificial Ethology*, 47–59.

6   See Holk Cruse, "Robotic Experiments on Insect Walking," in Holland and McFarland, eds., *Artificial Ethology*, 139.

7   "None of these results would have been obtained without the use of computer simulations," Cruse notes. "However, there remain important questions that could be answered only by hardware simulation, i.e. using a real robot. . . . Many behavioural properties are a function not of the brain alone, but of the brain, body and environment in combination. Thus we must somehow include the body and the environment in the simulation. (In our case this is relevant to the question of whether high-pass filtered positive feedback might really solve [certain kinds of] problems.) One of the best ways of doing this is to build a real robot." Cruse, "Robotic Experiments on Insect Walking," 139.

8   On this point, see Janet Wilde Astington, "The Developmental Interdependence of Theory of Mind and Language," in N. J. Enfield and Stephen C. Levinson, eds., *Roots of Sociality: Culture, Cognition and Interaction* (Oxford: Berg, 2006), 179–206; and Jennie E. Pyers, "Constructing the Social Mind: Language and False-Belief Understanding," in Enfield and Levinson, eds., *Roots of Sociality*, 207–228.

9   See Emmet Spier, "Robotic Experiments on Rat Instrumental Learning," in Holland and McFarland, eds., *Artificial Ethology,* 189–209; and Brendan McGonigle, "Robotic Experiments on Complexity and Cognition," in Holland and McFarland, eds., *Artificial Ethology,* 210–224.

10 See Francisco J. Varela, Evan Thompson, and Eleanor Rosch, *The Embodied Mind: Cognitive Science and Human Experience* (Cambridge, Mass.: MIT Press, 1991); see also Rolf Pfeifer and Alexandre Pitti, *La révolution de l'intelligence du corps* (Paris: Manuella, 2012).

11 See E. Thompson and F. J. Varela, "Radical embodiment: Neural dynamics and consciousness," *Trends in Cognitive Sciences* 5, no. 10 (2001): 418–425; and Anthony Chemero, *Radical Embodied Cognitive Science* (Cambridge, Mass.: MIT Press, 2010).

12 John Stewart, Olivier Gapenne, and Ezequiel A. Di Paolo, eds., *Enaction: Toward a New Paradigm for Cognitive Science* (Cambridge, Mass.: MIT Press, 2010).

13 See, for example, Cruse, "Robotic Experiments on Insect Walking," and McGonigle, "Robotic Experiments on Complexity and Cognition."

14 Étienne Bimbenet, in *L'animal que je ne suis plus* (Paris: Gallimard, 2011), proposes a philosophical analysis of the difference between animal and human minds according to which the former are taken to be highly local in just this sense by comparison with the latter.

15 This raises, by the way, a certain number of epistemological questions associated with extrapolation from one type of system to another; see Daniel P. Steel, *Across the Boundaries: Extrapolation in Biology and Social Science* (New York: Oxford University Press, 2008).

16 See Humberto R. Maturana and Francisco J. Varela, *Autopoiesis and Cognition: The Realization of the Living,* Boston Studies in Philosophy of Science, vol. 42 (Dordrecht: D. Reidel, 1980).

17 See Evan Thompson, *Mind in Life: Biology, Phenomenology, and Sciences of Mind* (Cambridge, Mass.: Belknap Press of Harvard University Press, 2007).

18 See M. Bitbol and P. L. Luisi, "Autopoiesis with or without cognition: Defining life at its edge," *Journal of the Royal Society Interface* 1, no. 1 (2004): 99–107.

19 See Humberto R. Maturana and Francisco J. Varela, *The Tree of Knowledge: The Biological Roots of Human Understanding,* trans. Robert Paolucci (Boston: Shambhala / New Science Library, 1987; rev. ed., 1992). This thesis occupies an important place in theories of self-organization and in biological theories of cognition concerned chiefly with the autonomy of living systems. It is explicitly affirmed in the founding text of the embodied mind school, Varela, Thompson, and Rosch's *The Embodied Mind,* and today remains implicit in the work of most

theorists who share their general approach, though some frankly reject it; only members of the radical embodiment movement readily grasp its significance.

20   Thus Descartes opposed the accepted view of his time, namely, that there exist three types of soul: a vegetative soul, an animal (or sensitive) soul, and a rational soul. For Descartes, the soul has but a single form, rational and spiritual, which is constitutive of human beings as conscious people; all the rest is mere matter.

21   Later, we will see that the apparent mechanization of the mind brought about by the computer is profoundly ambiguous, and indeed dualist, for what it essentially involves is a *mathematization* of the mind.

22   René Descartes, *Discourse on the Method for Guiding One's Reason* [AT 6:58–59]; in *A Discourse on Method and Related Writings*, trans. Desmond M. Clarke (London: Penguin, 1999), 42. Descartes later reaffirmed this position, albeit rather obliquely, in the *Meditations*, a highly philosophical and metaphysical work; see *Meditations on First Philosophy* [AT 7:87–88], in *Meditations and Other Metaphysical Writings*, trans. Desmond M. Clarke (London: Penguin, 1998), 66–67. Not long afterward, he returned to the problem of animal intelligence in the letter to Henry More (February 5, 1649); see Descartes, *Meditations and Other Metaphysical Writings*, 173–175.

23   Thierry Gontier, "Descartes et les animaux-machines: Une réhabilitation?" in Jean-Luc Guichet, ed., *De l'animal-machine à l'âme des machines: Querelles biomécaniques de l'âme, XVIIᵉ–XXIᵉ siècles* (Paris: Publications de la Sorbonne, 2010), 38.

24   See Jerry A. Fodor, *The Modularity of Mind: An Essay on Faculty Psychology* (Cambridge, Mass.: MIT Press, 1983).

25   See David J. Chalmers, *The Conscious Mind: In Search of a Fundamental Theory* (New York: Oxford University Press, 1996).

26   See Mark Rowlands, *The New Science of the Mind: From Extended Mind to Embodied Phenomenology* (Cambridge, Mass.: MIT Press, 2010).

27   Fred Adams and Ken Aizawa, "Defending the Bounds of Cognition," in Richard Menary, ed., *The Extended Mind* (Cambridge, Mass.: MIT Press, 2010), 67–80.

28   See A. Clark and D. J. Chalmers, "The extended mind," *Analysis* 58, no. 1 (1998): 10–23.

29   See Chalmers, *Conscious Mind*, especially 164–168; see also Jaegwon Kim, "The Myth of Nonreductive Materialism," in *Supervenience and Mind: Selected Philosophical Essays* (New York: Cambridge University Press, 1993), 265–284.

30   The empirical thesis has recently been developed in a very interesting fashion in the context of cognitive archaeology by Lambros Malafouris; for a summary of the material-engagement theory (MET), see Lambros Malafouris, *How Things*

Shape the Mind: A Theory of Material Engagement (Cambridge, Mass.: MIT Press, 2013), 23–53.

31 See Andy Clark, *Natural-Born Cyborgs: Minds, Technologies, and the Future of Human Intelligence* (New York: Oxford University Press, 2003). Following the 1998 article written with Chalmers, Clark became the chief advocate of this thesis in a number of books and articles.

32 See, in particular, the essays collected in Menary, ed., *Extended Mind*.

33 See Paul Humphreys, *Extending Ourselves: Computational Science, Empiricism, and Scientific Method* (New York: Oxford University Press, 2004).

34 Humphreys, *Extending Ourselves*, vii.

35 Humphreys, *Extending Ourselves*, 53.

36 See Johannes Lenhard and Eric Winsberg, "Holism and Entrenchment in Climate Model Validation," in Martin Carrier and Alfred Nordmann, eds., *Science in the Context of Application*, Boston Studies in the Philosophy of Science, vol. 274 (New York: Springer, 2011), 115–130.

37 See Denis Le Bihan, *Looking Inside the Brain*, trans. T. Lavender Fagan (Princeton, N.J.: Princeton University Press, 2015).

38 No one, that is, apart from a small number of radiologists who have expert knowledge of the technical characteristics of the device. I owe this qualification to a private conversation with Denis Le Bihan.

39 For a careful analysis of complex models in both the social sciences and in fundamental physics that very often are neither analytically transparent nor epistemically penetrable, so that we can neither entirely explain them nor wholly understand how they operate, see Mary S. Morgan and Margaret Morrison, eds., *Models as Mediators: Perspectives on Natural and Social Science* (Cambridge: Cambridge University Press, 1999).

40 See Rowlands, *New Science of the Mind*, 154.

41 See David Marr, *Vision: A Computational Investigation into the Human Representation and Processing of Visual Information* (San Francisco: W. H. Freeman, 1982).

42 Rowlands, *New Science of the Mind*, 155.

43 See Rowlands, *New Science of the Mind*, 107–134.

44 A genome sequencer is a device capable of automating the sequencing of DNA. The sequencer is used to determine the order of DNA nucleotide bases and to display it, after processing, in the form of a series of letters representing the individual nucleotides.

45 See Amartya Sen, *The Idea of Justice* (Cambridge, Mass.: Belknap Press of Harvard University Press, 2009), 225–321.

46 In this connection, see the U.S. government website http://www.gps.gov.

3. MIND, EMOTIONS, AND ARTIFICIAL EMPATHY

1 In this connection, see Hilary Putnam, "Psychological Predicates," in W. H. Cap-
itan and D. D. Merrill, eds., *Art, Mind, and Religion* (Pittsburgh: University of
Pittsburgh Press, 1967), 37–48; J. A. Fodor, "Special sciences (or: The disunity of
science as a working hypothesis," *Synthese* 28, no. 2 (1974): 97–115; Jerry A. Fodor,
The Language of Thought (New York: Thomas Crowell, 1975); L. A. Shapiro, "Mul-
tiple realizations," *Journal of Philosophy* 97, no. 12 (2000): 635–654; Lawrence A.
Shapiro, *The Mind Incarnate* (Cambridge, Mass.: MIT Press, 2004); and John
Bickle, "Multiple Realizability," in Edward N. Zalta, ed., *Stanford Encyclopedia of
Philosophy* (Spring 2013 ed.), http://plato.stanford.edu/archives/spr2013/entries
/multiple-realizability.

2 See Andy Clark, "*Memento's* Revenge: The Extended Mind, Extended," in
Richard Menary, ed., *The Extended Mind* (Cambridge, Mass.: MIT Press,
2010), 64.

3 Clark, "*Memento's* Revenge," 44.

4 This tradition played a fundamental role in the thought not only of Husserl,
but also of Wittgenstein. Nearer our own time, we might mention Donald Davidson, another important representative of the tradition in analytic philosophy. See Davidson's essay "Knowing one's own mind," *Proceedings and Addresses
of the American Philosophical Association* 60, no. 3 (1987): 441–458, reprinted in
Quassim Cassam, ed., *Self-Knowledge* (Oxford: Oxford University Press, 1994),
43–64.

5 See, for example, D. Baron and D. C. Long, "Avowals and first-person privilege,"
Philosophy and Phenomenological Research 62, no. 2 (2001): 311–335.

6 As this prejudice is at the heart of the phenomenological approach inaugurated
by Husserl, the chance that it might escape the clutches of solipsism seems slight.

7 This hypothesis remains entrenched in cognitive psychology today. Theory of
mind, taking as its point of departure the idea that we do not have direct access
to the minds of others, holds that to attribute to them a mind, a set of mental
states, is not the result of observation, but a mere hypothesis, a theory. This
branch of inquiry is known as the theory-theory approach (sometimes also called
"theory of the theory of mind"). See, for example, Peter Carruthers and Peter K.
Smith, eds., *Theories of Theories of Mind* (Cambridge: Cambridge University
Press, 1996).

8 Note that this example is not to be found in the Aristotelian corpus itself, but in
Sextus Empiricus, *Outlines of Pyrrhonism*, 1.14.

9 Descartes, *Meditations on First Philosophy* [AT 7.22]; in *Meditations and Other
Metaphysical Writings*, 19.

10 See K. Dautenhahn, "Methodology and themes of human-robot interaction: A growing research field," *International Journal of Advanced Robotics* 4, no. 1 (2007): 103–108.

11 See Cynthia Breazeal, "Artificial Interaction between Humans and Robots," in Jozef Kelemen and Petr Sosik, eds., *Advances in Artificial Life (ECAL 2001)*, Lecture Notes in Computer Science, vol. 2159 (Heidelberg: Springer, 2001), 582–591.

12 See Hector Levesque and Gerhard Lakemeyer, "Cognitive Robotics," in Frank van Harmelen, Vladimir Lifschitz, and Bruce Porter, eds., *Handbook of Knowledge Representation* (Amsterdam: Elsevier, 2008), 869–886.

13 See M. Asada, "Towards artificial empathy: How can artificial empathy follow the developmental pathway of natural empathy?," *International Journal of Social Robotics* 7, no. 1 (2015): 19–33.

14 See L. Berthouze and T. Ziemke, "Epigenetic robotics: Modeling cognitive development in robotic systems," *Connection Science* 15, no. 3 (2003): 147–150.

15 David Feil-Seifer and Maja J. Matarić, "Defining Socially Assistive Robotics," in *Proceedings of the 2005 IEEE 9th International Conference on Rehabilitation Robotics [ICORR]* (Piscataway, N.J.: Institute of Electrical and Electronics Engineers, 2005), 465–468.

16 See Michael Hillman, "Rehabilitation Robotics from Past to Present—A Historical Perspective," in Z. Zenn Bien and Dimitar Stefanov, eds., *Advances in Rehabilitation Robotics: Human-Friendly Technologies on Movement Assistance and Restoration for People with Disabilities* (Berlin: Springer, 2004), 25–44.

17 The technical literature contains several definitions of the kind of social robot that we will have occasion to meet in this chapter. All refer to a class of robots for which social interaction plays a key role; see T. Fong, I. Nourbakhsh, and K. Dautenhahn, "A survey of socially interactive robots," *Robotics and Autonomous Systems* 42, nos. 3–4 (2003): 146–166.

18 M. Heerink, B. Kröse, V. Evers, and B. J. Wielinga, "The influence of social presence on acceptance of a companion robot by older people," *Journal of Physical Agents* 2, no. 2 (2008): 33–40, emphasis added. This notion comes from the field of human-computer interaction (HCI), where the need to create "affective interfaces" was first perceived. See also Ana Paiva, ed., *Affective Interactions: Towards a New Generation of Computer Interfaces*, Lecture Notes in Computer Science / Lecture Notes in Artificial Intelligence, vol. 1814 (Berlin: Springer, 2000).

19 Emphasis on the need to take affective abilities into account in designing social robots is now a commonplace of the robotics literature. See, for example, Fong et al., "Survey of socially interactive robots"; and H. Li, J.-J. Cabibihan, and Y. K. Tan, "Towards an effective design of social robots," *International Journal of Social Robotics*

3, no. 4 (2011): 333–335. On the importance of the affective dimension in interactions with artificial agents, see Paiva, ed., *Affective Interactions;* also Ana Paiva, Iolanda Leite, and Tiago Ribeiro, "Emotion Modeling for Social Robots," in Rafael A. Calvo, Sidney K. D'Mello, Jonathan Gratch, and Arvid Kappas, eds., *Oxford Handbook of Affective Computing* (Oxford: Oxford University Press, 2014), 296–308.

20 The concept of social acceptance, in relation to technology in general, is conceived as "the demonstrable willingness within a user group to employ information technology for the tasks it is designed to support"; see A. Dillon and M. G. Morris, "User acceptance of information technology: Theories and models," *Annual Review of Information Science and Technology* 31 (1996): 4. See, too, Ioannis Vasmatzidis, "User Requirements in Information Technology," in Waldemar Karwowski, ed., *International Encyclopedia of Ergonomics and Human Factors,* 3 vols. (London: Taylor and Francis, 2001), 1:750–753. In connection with social robots, the notion refers to the ability of artificial agents to respond to human social expectations on a variety of levels: aesthetic (a pleasing manner), communicative (comprehension of verbal and nonverbal communication), behavioral (respect for social rules), and affective (generating positive emotions in human partners and creating believable affective relationships). See Jenay M. Beer, Akanksha Prakash, Tracy L. Mitzner, and Wendy A. Rogers, "Understanding Robot Acceptance," Technical Report HFA-TR-1103 (Atlanta: Georgia Institute of Technology, School of Psychology, 2011), https://smartech.gatech.edu/bitstream/handle/1853/39672/HFA-TR-1103-RobotAcceptance.pdf; and Kerstin Dautenhahn, "Design Spaces and Niche Spaces of Believable Social Robots," in *Proceedings of the 11th IEEE International Workshop on Robot and Human Interactive Communication* (New York: Institute of Electrical and Electronics Engineers, 2002), 192–197.

21 See Rafael A. Calvo, Sidney K. D'Mello, Jonathan Gratch, and Arvid Kappas, "Introduction to Affective Computing," in Calvo et al., eds., *Oxford Handbook of Affective Computing,* 1–8.

22 A point made earlier, at the end of Chapter 2, in connection with Otto's hailing a taxi to go to the Museum of Modern Art.

23 See Gregory Bateson, *Steps to an Ecology of Mind* (San Francisco: Chandler / New York: Ballantine, 1972).

24 See M. A. Arbib and J. M. Fellous, "Emotions: From brain to robot," *Trends in Cognitive Sciences* 8, no. 12 (2004): 554–561.

25 See Arbib and Fellous, "Emotions," 554; see also T. Ziemke and R. Lowe, "On the role of emotion in embodied cognitive architectures: From organisms to robots," *Cognitive Computation* 1, no. 1 (2009): 104–117.

26 See Jean Piaget, *The Child's Conception of the World*, trans. Joan Tomlinson and Andrew Tomlinson (London: Routledge and Kegan Paul, 1929).

27 For a critical survey of recent research on anthropomorphism, see G. Airenti, "Aux origines de l'anthropomorphisme: Intersubjectivité et théorie de l'esprit," *Gradhiva* 15 (2012): 35–53.

28 Owing to the close links that exist today between social robotics, the arts, and psychology, this conjecture has also acquired a growing importance in disciplines that at first glance seem to have very little in common with robotics.

29 See, for example, Kerstin Dautenhahn, "Robots as Social Actors: AURORA and the Case of Autism," in *Proceedings of the Third International Cognitive Technology Conference* (San Francisco, August 1999), 359–374.

30 If the rising curve of familiarity and ease of interaction suddenly plunges into the valley when a robot resembles us too much, this, we believe, is because the robot has become an "other"—an unknown agent whose reactions cannot be anticipated outside, or even within, a shared cultural framework. What is at issue in the collapse of familiarity is therefore just this: that too perfect a resemblance modifies the characteristics of the interaction.

31 See C. Breazeal, "Toward sociable robots," *Robotics and Autonomous Systems* 42, nos. 3–4 (2003): 167–175.

32 Fong et al., "Survey of socially interactive robots," 145.

33 Fong et al., "Survey of socially interactive robots," 146.

34 Paiva et al., "Emotion Modeling for Social Robots," 297.

35 See Rafael Núñez and Walter J. Freeman, eds., *Reclaiming Cognition: The Primacy of Action, Intention, and Emotion* (Thorverton, Devon, U.K.: Imprint Academic, 2000); see also Luisa Damiano, *Unità in dialogo: Un nuovo stile per la conoscenza* (Milan: Bruno Mondadori, 2009).

36 See E. A. Phelps, "Emotion and cognition: Insights from studies of the human amygdala," *Annual Review of Psychology* 57 (2006): 27–53.

37 See Tom Ziemke, "What's That Thing Called Embodiment?" in Richard Alterman and David Kirsh, eds., *Proceedings of the 25th Annual Meeting of the Cognitive Society* (Boston, Mass.: Cognitive Science Society, 2003), 1305–1310.

38 See A. Clark, "An embodied cognitive science?," *Trends in Cognitive Sciences* 3, no. 9 (1999): 345–351; see also Raymond W. Gibbs, Jr., *Embodiment and Cognitive Science* (New York: Cambridge University Press, 2000).

39 See Ziemke, "What's That Thing Called Embodiment?"

40 See Rolf Pfeifer and Christian Scheier, *Understanding Intelligence* (Cambridge, Mass.: MIT Press, 1999).

41 The experiments by researchers in artificial ethology mentioned earlier, in Chapter 2, illustrate this approach.

42 It has long been held, at least since Vico and the advent of modern science, that one really understands only that which one is capable of making oneself. See Luisa Damiano and Lola Cañamero, "Constructing Emotions: Epistemological Groundings and Applications in Robotics for a Synthetic Approach to Emotions," in Jackie Chapell, Susannah Thorpe, Nick Hawes, and Aaron Sloman, eds., *Proceedings of the International Symposium on AI-Inspired Biology 2010* (London: Society for the Study of Artificial Intelligence and the Simulation of Behaviour, 2010), 20–28, http://www.cs.bham.ac.uk/~nah/bibtex/papers/aiib -proceedings.pdf.

43 See Damiano and Cañamero, "Constructing Emotions."

44 See Valentino Braitenberg, *Vehicles: Experiments in Synthetic Psychology* (Cambridge, Mass.: MIT Press, 1984).

45 See L. Cañamero, "Emotion understanding from the perspective of autonomous robots research," *Neural Networks* 18, no. 4 (2005): 445–455.

46 Ziemke and Lowe, "Emotion in embodied cognitive architectures," 108, emphasis added.

47 Ziemke and Lowe, "Emotion in embodied cognitive architectures," 108.

48 Domenico Parisi, *Future Robots: Towards a Robotic Science of Human Beings* (London: John Benjamins, 2014), 69, emphasis in the original. See also D. Parisi, "Internal robotics," *Connection Science* 16, no. 4 (2004): 325–338.

49 Ziemke and Lowe, for example, have proposed an architecture consisting of three different levels of homeostatic processes regulating behavioral organization; see "Emotion in embodied cognitive architectures."

50 Almost fifteen years ago, the mesolimbic dopamine system of the mammalian brain had already been modeled by the robotic platform MONAD; see W. H. Alexander and O. Sporns, "An embodied model of learning, plasticity, and reward," *Adaptive Behavior* 10, nos. 3–4 (2002): 143–159.

51 One example is the affective developmental robotics approach developed by Minoru Asada, building on the principles of developmental cognitive robotics to deepen our intuitions about self / other knowledge; see Asada, "Towards artificial empathy."

52 Ralph Adolphs, "Could a Robot Have Emotions? Theoretical Perspectives from Social Cognitive Neuroscience," in Jean-Marc Fellous and Michael A. Arbib, eds., *Who Needs Emotions? The Brain Meets the Robot* (New York: Oxford University Press, 2005), 9.

53   On social intelligence, see Fong et al., "Survey of socially interactive robots." This approach faces considerable difficulties, not least in regard to the many methodological questions associated with the level of abstraction appropriate to a synthetic model and the degree of biological detail that a good model must have. On this last point, see chapters 5 and 8 in Daniel P. Steel, *Across the Boundaries: Extrapolation in Biology and Social Science* (New York: Oxford University Press, 2008), 78–100, 149–173.

54   See Thomas Dixon, *From Passions to Emotions: The Creation of a Secular Psychological Category* (Cambridge: Cambridge University Press, 2003), especially 98–134.

55   See Braitenberg, *Vehicles*.

56   See K. Höök, "Affective loop experiences: Designing for interactional embodiment," *Philosophical Transactions of the Royal Society B: Biological Sciences* 364, no. 12 (2009): 3585–3595. See also Paiva et al., "Emotion Modeling for Social Robots"; Fong et al., "Survey of socially interactive robots"; and L. Cañamero, "Bridging the Gap between HRI and Neuroscience in Emotion Research: Robots as Models," paper presented at the 2014 Human-Robot Interaction [HRI] Workshop, Bielefeld, Germany, http://www.macs.hw.ac.uk/~kl360/HRI2014W/submission/S16.pdf.

57   Kristina Höök, "Affective Loop Experiences—What Are They?," in Harri Oinas-Kukkonen, Per Hasle, Marja Harjumaa, Katarina Segerståhl, and Peter Øhrstrøm, eds., *Persuasive Technology: Proceedings of Third International Conference*, Lecture Notes in Computer Science, vol. 5033 (Berlin: Springer, 2008), 2. [Punctuation and grammar slightly modified.—Trans.]

58   Paiva et al., "Emotion Modeling for Social Robots," 297.

59   See Frank Hegel, Torsten Spexard, Britta Wrede, Gernot Horstmann, and Thurid Vogt, "Playing a Different Imitation Game: Interaction with an Empathetic Android Robot," in *Sixth IEEE-RAS International Conference on Humanoid Robots* (Piscataway, N.J.: Institute of Electrical and Electronics Engineers, 2006), 56–61.

60   See C. Breazeal, "Emotions and sociable humanoid robots," *International Journal of Human-Computer Studies* 59, no. 1 (2003): 119–155.

4. THE OTHER OTHERWISE

1   See Thomas Hobbes, *Human Nature; or, The Fundamental Elements of Policy* (1650), 9.1–21, part 1 of *The Elements of Law, Natural and Politic*, ed. J. C. A. Gaskin (Oxford: Oxford University Press, 1994), 50–60. For a more recent and more complete version, see Paul Dumouchel, *Émotions: Essai sur le corps et le social* (Paris: Les Empêcheurs de Penser en Rond, 1999), 61–70; see also Sergio Manghi's preface, "Legame emozionale, legame sociale," to the Italian edition of the

same book, *Emozioni: Saggio sul corpo e sul sociale,* trans. Luisa Damiano (Milan: Medusa, 2008), 5–14.

2 The expression "mirror mechanisms," attributed to Vittorio Gallese, refers in a general way to mirror neurons, found in area F5 of the macaque premotor cortex, as well as to the mirror neuron systems found in humans. See G. Rizzolatti, L. Fogassi, and V. Gallese, "Neurological mechanisms underlying the understanding and imitation of action," *Nature Reviews Neuroscience* 2, no. 9 (2001): 661–670.

3 See Dumouchel, *Émotions,* 83–108; see also Luisa Damiano and Paul Dumouchel, "Epigenetic Embodiment," in Lola Cañamero, Pierre-Yves Oudeyer, and Christian Balkenius, eds., *Proceedings of the Ninth International Conference on Epigenetic Robotics,* Lund University Cognitive Studies, vol. 146 (Lund: Department of Philosophy, Lund University, 2009), 41–48.

4 See Paul Ekman and Wallace V. Friesen, *Unmasking the Face: A Guide to Recognizing Emotions from Facial Clues* (Englewood Cliffs, N.J.: Prentice-Hall, 1975).

5 See Carroll E. Izard, "Cross-Cultural Perspectives on Emotion and Emotion Communication," in Harry C. Triandis and Walter J. Lonner, eds., *Basic Processes,* vol. 3 of *Handbook of Cross-Cultural Psychology* (Boston: Allyn and Bacon, 1980), 185–220.

6 See Don Ross and Paul Dumouchel, "Emotions as strategic signals," *Rationality and Society* 16, no. 3 (2004): 251–286; see also Allan Gibbard, *Wise Choices, Apt Feelings: A Theory of Normative Judgment* (Cambridge, Mass.: Harvard University Press, 1990).

7 See V. Gallese, "Intentional attunement: A neurophysiological perspective on social cognition and its disruption in autism," *Brain Research: Cognitive Brain Research* 1079, no. 1 (2006): 15–24; see also Vittorio Gallese, "The 'Shared Manifold' Hypothesis: Embodied Simulation and Its Role in Empathy and Social Cognition," in Tom F. D. Farrow and Peter W. R. Woodruff, eds., *Empathy in Mental Illness* (Cambridge: Cambridge University Press, 2007), 448–472.

8 See Vittorio Gallese, " 'Being Like Me': Self-Other Identity, Mirror Neurons, and Empathy," in Susan Hurley and Nick Chater, eds., *Perspectives on Imitation: From Neuroscience to Social Science,* 2 vols. (Cambridge, Mass.: MIT Press, 2005), 1:101–118.

9 In addition to recent results in neuroscience and cognitive science that support the hypothesis of affective coordination, see the essays in Marc D. Lewis and Isabela Granic, eds., *Emotion, Development, and Self-Organization: Dynamic Systems Approaches to Emotional Development* (New York: Cambridge University Press, 2002); and Raymond W. Gibbs, Jr., "Intentions as Emergent Products of Social Interactions," in Bertram F. Malle, Louis J. Moses, and Dare A. Baldwin, eds., *In-*

tentions and Intentionality: Foundations of Social Cognition (Cambridge, Mass.: MIT Press, 2001), 105–122. It is interesting that this new approach to affective phenomena should rediscover certain landmarks in the intellectual genealogy of embodied cognitive science, such as the thesis of the interindividual regulation of action (see Jean Piaget, *Biology and Knowledge: An Essay on the Relations between Organic Regulations and Cognitive Processes,* trans. Beatrix Walsh [Chicago: University of Chicago Press, 1971]), the autopoietic theory of behavioral coupling (see Humberto R. Maturana and Francisco J. Varela, *The Tree of Knowledge: The Biological Roots of Human Understanding,* trans. Robert Paolucci [Boston: Shambhala / New Science Library, 1987; rev. ed., 1992]), and the idea of conversational unity introduced by Gordon Pask (in *Conversation, Cognition, and Learning: A Cybernetic Theory and Methodology* [Amsterdam: Elsevier, 1975]) and subsequently developed by Francisco Varela (in *Principles of Biological Autonomy* [New York: North Holland, 1979]). See also part II of Luisa Damiano, *Unità in dialogo: Un nuovo stile per la conoscenza* (Milan: Bruno Mondadori, 2009).

10 See Dumouchel, *Émotions,* 115–121.

11 When Paul Dumouchel was little, his family had a dog that scratched at the kitchen door when it wanted to go out. Paul's elderly great aunt, who was reading in the living room, would slowly get up from her comfortable armchair and go open the door. When she got to the kitchen, the dog was no longer there. It had gone around through another room in order to avoid meeting her, and when she came back to the living room she found it happily occupying her chair.

12 See Luisa Damiano, Paul Dumouchel, and Hagen Lehmann, "Should Empathic Social Robots Have Interiority?," in Shuzhi Sam Ge, Oussama Khatib, John-John Cabibihan, Reid Simmons, and Mary-Anne Williams, eds., *Social Robotics: Fourth International Conference [ICSR 2012],* Lecture Notes in Artificial Intelligence, vol. 7621 (Berlin: Springer, 2012), 268–277; also L. Damiano, P. Dumouchel, and H. Lehmann, "Towards human-robot affective co-evolution: Overcoming oppositions in constructing emotions and empathy," *International Journal of Social Robotics* 7, no. 1 (2015): 7–18.

13 See, for example, Gibbs, "Intentions as Emergent Products of Social Interactions"; J. Krueger, "Extended cognition and the space of social interaction," *Consciousness and Cognition* 20, no. 3 (2011): 643–657; and Jan Slaby, "Emotions and Extended Mind," in Christian von Scheve and Mikko Salmela, eds., *Collective Emotions* (Oxford: Oxford University Press, 2014), 32–46.

14 See M. Ceruti and L. Damiano, "Embodiment enattivo, cognizione e relazione dialogica," *Encyclopaideia* 37, no. 17 (2013): 19–46.

15   See Francisco J. Varela, Evan Thompson, and Eleanor Rosch, *The Embodied Mind: Cognitive Science and Human Experience* (Cambridge, Mass.: MIT Press, 1991), 147–184; also E. Thompson and F. J. Varela, "Radical embodiment: Neural dynamics and consciousness," *Trends in Cognitive Sciences* 5, no. 10 (2001): 418–425.

16   See Damiano, *Unità in dialogo,* and the articles collected in John Stewart, Olivier Gapenne, and Ezequiel A. Di Paolo, eds., *Enaction: Toward a New Paradigm for Cognitive Science* (Cambridge, Mass.: MIT Press, 2010). Even so, some exponents of radical embodiment continue to rely on spatial conceptualization and the metaphors typical of the extended mind approach. See, for example, Alva Noë, *Out of Our Heads: Why You Are Not Your Brain, and Other Lessons from the Biology of Consciousness* (New York: Hill and Wang, 2009); see also M. Silberstein and A. Chemero, "Complexity and extended phenomenological-cognitive systems," *Topics in Cognitive Science* 4, no. 1 (2012): 35–50. On the difference between spatial conceptualization and dynamic models of the embodiment of mind, see Damiano et al., "Should Empathic Social Robots Have Interiority?"

17   See H. J. Chiel and R. D. Beers, "The brain has a body: Adaptive behavior emerges from interactions of nervous system, body, and environment," *Trends in Neurosciences* 20, no. 12 (1997): 553–557; Giulio Sandini, Giorgio Metta, and David Vernon, "The *iCub* Cognitive Humanoid Robot: An Open-System Research Platform for Enactive Cognition," in Max Lungarella, Fumiya Iida, Josh Bongard, and Rolf Pfeifer, eds., *50 Years of Artificial Intelligence,* Lecture Notes in Computer Science, vol. 4850 (Berlin: Springer, 2007), 358–369; and T. Froese and T. Ziemke, "Enactive artificial intelligence: Investigating the systemic organization of life and mind," *Artificial Intelligence* 173, nos. 3–4 (2009): 466–500.

18   On the physiological consequences of affective interactions, see C. Sue Carter, I. Izja Lederhendler, and Brian Kirkpatrick, eds., *The Integrative Neurobiology of Affiliation* (Cambridge, Mass.: MIT Press, 1999), especially chapters 5, 7–8, 14–16; see also Paul Dumouchel, "Agency, Affiliation, and Emotion," in Robert Trappl, ed., *Cybernetics and Systems: Proceedings of 11th European Meeting on Cybernetics and Systems Research [EMCSR]* (Vienna: Austrian Society for Cybernetics Studies, 2002), 727–732.

19   A view explicitly promoted, for example, by Ralph Adolphs, "Could a Robot Have Emotions? Theoretical Perspectives from Social Cognitive Neuroscience," in Jean-Marc Fellous and Michael A. Arbib, eds., *Who Needs Emotions? The Brain Meets the Robot* (New York: Oxford University Press, 2005), 9–25. See also Ana Paiva, Iolanda Leite, and Tiago Ribeiro, "Emotion Modeling for Social Robots," in Rafael A. Calvo, Sidney K. D'Mello, Jonathan Gratch, and Arvid Kappas, eds.,

Oxford Handbook of Affective Computing (Oxford: Oxford University Press, 2014), 296–308; D. Parisi, "Internal robotics," *Connection Science* 16, no. 4 (2004): 325–338; and Domenico Parisi, *Future Robots: Towards a Robotic Science of Human Beings* (London: John Benjamins, 2014).

20 See, for example, Chrystopher L. Nehaniv and Kerstin Dautenhahn, eds., *Imitation and Social Learning in Robots, Humans, and Animals: Behavioural, Social, and Communicative Dimensions* (Cambridge: Cambridge University Press, 2007); see also Kerstin Dautenhahn and Chrystopher L. Nehaniv, eds., *Imitation in Animals and Artifacts* (Cambridge, Mass.: MIT Press, 2002).

21 "ATR" is an abbreviation for Advanced Telecommunication Research Institute International, located in Kansai Science City between Kyoto, Osaka, and Nara. See http://www.atr.jp/index_e.html.

22 The experiments are described and analyzed in Emmanuel Grimaud and Zaven Paré, *Le jour où les robots mangeront des pommes: Conversations avec un Geminoïd* (Paris: Petra, 2011).

23 Obviously, the effect of social presence in this case has other components than the robot's gaze, but in the experiments conducted by Paré and Straub, this is the most important aspect.

24 Knowing that an impression is false and that the robot is not really looking at us puts one in mind of the Müller-Lyer illusion, where knowing that the two lines are identical in length does not prevent their being seen as different in length.

25 See Takanori Shibata, Kazuyoshi Wada, Tomoko Saito, and Kazuo Tanie, "Psychological and Social Effects to Elderly People by Robot-Assisted Activity," in Lola Cañamero and Ruth Aylett, eds., *Animating Expressive Characters for Social Interaction* (Amsterdam: John Benjamins, 2008), 177–193. Paro was briefly introduced earlier, in Chapter 1.

26 Its architecture comprises two hierarchical processing layers that generate different kinds of behavior. "The behavior generation layer," Shibata notes, "generates control references for each actuator to perform the determined behavior. The control reference depends on the magnitude of the internal states and their variation. For example, parameters can change the speed of movements or the number of instances of the same behavior. Therefore although the number of basic patterns is finite, the number of emerging behaviors is infinite because of the varying number of parameters. This creates life-like behavior." Shibata et al., "Psychological and Social Effects to Elderly People," 184.

27 See Shibata et al., "Psychological and Social Effects to Elderly People," 184.

28 These effects may in fact last for several years, but there is no way to be sure of this on the basis of current studies. See K. Wada and T. Shibata, "Living with seal

robots—Its sociopsychological and physiological influences on the elderly at a care house," *IEEE Transactions on Robotics* 23, no. 5 (2007): 972–980; also K. Wada, T. Shibata, M. Asada, and T. Musha, "Robot therapy for prevention of dementia at home—Results of preliminary experiment," *Journal of Robotics and Mechatronics* 19, no. 6 (2007): 691–697.

29   The current cost of Paro is about five thousand dollars; see A. Burton, "Dolphins, dogs, and robot seals for the treatment of neurological diseases," *Lancet Neurology* 12, no. 9 (2013): 851–852.

30   This is often but not always the case. As the Danish filmmaker Phie Ambo's documentary *Mechanical Love* (2007) shows, it sometimes happens that Paro becomes the occasion of conflict and that a person isolates himself or herself with it, avoiding common rooms where he or she is no longer welcome. See the short trailer for the film and related information at http://danishdocumentary.muvies .com/reviews/180-mechanical-love.

31   The project later migrated with Dautenhahn to the University of Hertfordshire. See http://www.herts.ac.uk/kaspar/the-social-robot.

32   See K. Dautenhahn, "Socially intelligent robots: Dimensions of human-robot interaction," *Philosophical Transactions of the Royal Society B: Biological Sciences* 362, no. 1480 (2007): 679–704; and K. Dautenhahn, C. L. Nehaniv, M. L. Walters, B. Robins, H. Kose-Bagci, N. A. Mirza, and M. Blow, "KASPAR—A minimally expressive humanoid robot for human-robot interaction research," *Applied Bionics and Biomechanics* 6, nos. 3–4 (2009): 369–397.

33   It is interesting and significant that the minimal expressiveness that makes KASPAR an appealing interaction partner for autistic children makes it a rather disturbing object for other people, particularly adults.

34   See L. D. Riek, "Wizard of Oz studies in HRI: A systematic review and new reporting guidelines," *Journal of Human-Robot Interaction* 1, no. 1 (2012): 119–136.

35   See Hagen Lehmann, Iolanda Iacono, Ben Robins, Patrizia Marti, and Kerstin Dautenhahn, " 'Make It Move': Playing Cause and Effect Games with a Robot Companion for Children with Cognitive Disabilities," in *Proceedings of the 29th Annual European Conference on Cognitive Ergonomics [ECCE 2011]* (New York: Association for Computing Machinery, 2011), 105–112, http://dl.acm.org/citation .cfm?id=2074712.2074734; H. Lehmann, I. Iacono, K. Dautenhahn, P. Marti, and B. Robins, "Robot companions for children with Down Syndrome: A case study," *Interaction Studies* 15, no. 1 (2014): 99–112; Iolanda Iacono, Hagen Lehmann, Patrizia Marti, Ben Robins, and Kerstin Dautenhahn, "Robots as Social Mediators for Children with Autism—A Preliminary Analysis Comparing Two Different Robotic Platforms," in *IEEE International Conference on Development*

and Learning [ICDL 2011], 2 vols. (Piscataway, N.J.: Institute of Electrical and Electronics Engineers, 2011), 2:1–6; Sandra Costa, Hagen Lehmann, Ben Robins, Kersten Dautenhahn, and Filomena Soares, "Where Is Your Nose? Developing Body Awareness Skills among Children with Autism Using a Humanoid Robot," in *Proceedings of the Sixth International Conference on Advances in Computer-Human Interactions [ACHI 2013]* (Copenhagen: IARIA XPS Press, 2013), 117–122; S. Costa, H. Lehmann, K. Dautenhahn, B. Robins, and F. Soares, "Using a humanoid robot to elicit body awareness and appropriate physical interaction in children with autism," *International Journal of Social Robotics* 6, no. 4 (2014): 1–14; and J. Wainer, K. Dautenhahn, B. Robins, and F. Amirabdollahian, "A pilot study with a novel setup for collaborative play of the humanoid robot KASPAR with children with autism," *International Journal of Social Robotics* 6, no. 1 (2014): 45–65.

36   See Aude Billard, Annalisa Bonfiglio, Giorgio Cannata, Piero Cosseddu, Torbjorn Dahl, Kerstin Dautenhahn, Fulvio Mastrogiovanni et al., "The RoboSkin Project: Challenges and Results," in Vincent Padois, Philippe Bidaud, and Oussama Khatib, eds., *Romansy 19—Robot Design, Dynamics, and Control*, CISM International Centre for Mechanical Sciences Series, vol. 544 (Vienna: Springer, 2013), 351–358.

37   See Luisa Damiano, *Filosofia della scienza e medicina riabilitativa in dialogo* (forthcoming).

38   The purchase price of KASPAR, deliberately kept as low as possible, is currently below 1,200 euros (about $1,350). Moreover, maintenance costs will soon be lower thanks to the development of 3D printers that make it possible for users to manufacture some replacement parts themselves, such as the hands, feet, certain parts of the head, and so on. This should make the technology more affordable for schools and specialized services for children with various learning disabilities. We may therefore expect to see a growing use of robots in these contexts, which in turn will force us to confront the particular ethical problems raised by such use.

39   See especially the section entitled "Intimacy" in Grimaud and Paré, *Le jour où les robots mangeront des pommes*, 75–78.

40   In this connection, see the recent film written and directed by Alex Garland, *Ex Machina* (2015).

41   To name only a few: bioethics, medical ethics, nanoethics, business ethics, robotic ethics, machine ethics, health care ethics, Darwinian ethics, feminist ethics, research ethics. This multiplication of ethical domains suggests, and sometimes takes for granted or even explicitly assumes, that the moral questions involved are radically different from one domain to the other, which is far from obvious.

42 See Hannah Arendt, *The Human Condition* (Chicago: University of Chicago Press, 1958), 7–8.

5. FROM MORAL AND LETHAL MACHINES TO SYNTHETIC ETHICS

1 This will also give us an opportunity once again to illustrate the diversity of the cognitive domain and to emphasize how different from one another the various artificial agents that we already know how to make can be.

2 In this sense, they resemble the simple machines of Bruno Latour's action programs, such as the turnspit and door key discussed in Chapter 1.

3 See Patrick Lin, Keith Abney, and George A. Bekey, eds., *Robot Ethics: The Ethical and Social Implications of Robotics* (Cambridge, Mass.: MIT Press, 2012).

4 See Wendell Wallach and Colin Allen, *Moral Machines: Teaching Robots Right from Wrong* (New York: Oxford University Press, 2008).

5 See, for example, M. Anderson and S. L. Anderson, "Machine ethics: Creating an ethical intelligent agent," *AI Magazine* 28, no. 4 (2007): 15–26; and J. H. Moor, "The nature, importance, and difficulty of machine ethics," *IEEE Intelligent Systems* 21, no. 4 (2006): 18–21.

6 See, for example, L. Floridi and J. W. Sanders, "On the morality of artificial agents," *Minds and Machines* 14, no. 3 (2004): 349–379, which proposes that artificial agents be subject to the same professional code of ethics as engineers. For Floridi and Sanders, however, it is not a matter of treating this code as a set of rules for determining proper behavior by artificial agents, but of using it as a criterion for evaluating their performance. We will come back to this issue later.

7 "Robot ethics" is not to be confused with "roboethics," a term that is sometimes used, in contrast to machine ethics, to denote the attempt to determine what counts as appropriate behavior for human beings in their relations with robots and other artificial agents.

8 Thus the subtitle of Wallach and Allen's book: *Teaching Robots Right from Wrong*.

9 See J. B. Schneewind, *The Invention of Autonomy: A History of Modern Moral Philosophy* (Cambridge: Cambridge University Press, 1998).

10 See Wallach and Allen, *Moral Machines*, especially chapter 2 ("Engineering Morality"), 25–36.

11 In this respect, its purpose resembles that of both the ethics of artificial life and of nanotechnologies to the extent that these ethics bear essentially on technological developments that *might* take place in the future. Concerning artificial life, see Mark A. Bedau and Emily C. Parke, eds., *The Ethics of Protocells: Moral and Social Implications of Creating Life in the Laboratory* (Cambridge, Mass.: MIT

Press, 2009); concerning nanotechnologies, see Dónal P. O'Mathúna, *Nanoethics: Big Ethical Issues with Small Technology* (London: Continuum, 2009).

12 Since it has long been common practice in various versions of utilitarian doctrine to quantify the moral value of different options for the purpose of comparing them, utilitarianism plainly recommends itself to anyone seeking to thoroughly "mechanize" moral reasoning.

13 Except perhaps in certain versions of utilitarianism. This requirement seems otherwise to be insisted upon by modern moral philosophers, probably including Bentham and certainly Mill.

14 Robot ethics is not alone in this. Our argument may readily be extended to other regional ethics. Here, however, the divergence arises from the fact that the impossibility of acting otherwise is an integral feature of the machine's internal structure, rather than of the structure of the external environment within which the artificial agent acts.

15 In this latter connection, see Schneewind, *Invention of Autonomy*, especially 4.22, 483–507; see also Paul Dumouchel, *The Barren Sacrifice: An Essay on Political Violence*, trans. Mary Baker (East Lansing: Michigan State University Press, 2015).

16 See Armin Krishnan, *Killer Robots: Legality and Ethicality of Autonomous Weapons* (Burlington, Vt.: Ashgate, 2009), 7.

17 See Krishnan, *Killer Robots*, 3.

18 See Krishnan, *Killer Robots*, 4.

19 Accordingly, just as science is now in large part essentially made by machines, as we saw in Chapter 2, so too military strategy is made, and will be made more and more, by machines.

20 This is also the opinion of two Chinese army strategists, Qiao Liang and Wang Xiangsui, authors of an important book published in English as *Unrestricted Warfare: China's Master Plan to Destroy America* (Panama City, Panama: Pan American, 2002), an abridged translation of the edition published in Beijing by the People's Liberation Army Literature and Arts Publishing House in 1999.

21 The first of these is Ronald C. Arkin, *Behavior-Based Robotics* (Cambridge, Mass.: MIT Press, 1998). This is the inaugural volume of the Intelligent Robotics and Autonomous Agents series, edited by Arkin. The fact that Arkin appears to be unaware of Krishnan's work in no way affects the argument we make here.

22 In fact, this is not quite accurate. Arkin already has an answer to this question that he finds satisfactory. His real problem, which, as we will see in what follows, is technical rather than ethical in nature, is how to make robots conform to rules that determine when killing is legitimate.

23 Here Arkin stands opposed to Krishnan, who, while he likewise considers this tendency to be inevitable, nevertheless does not consider legitimate the attempt to bring about such a state of affairs.

24 See Ronald C. Arkin, *Governing Lethal Behavior in Autonomous Robots* (Boca Raton, Fla.: CRC Press, 2009).

25 As a result, Arkin avoids a major question in robot ethics having to do with the justification of the ethical codes built into a machine's behavior. Given that these robots are, in effect, soldiers of the U.S. armed forces, the choice that has been made with regard to laws of war and of American military norms seems obvious to him and in any case immune to challenge.

26 See Arkin, *Governing Lethal Behavior in Autonomous Robots*, 29–30.

27 See Arkin, *Governing Lethal Behavior in Autonomous Robots*, 135.

28 Arkin, *Governing Lethal Behavior in Autonomous Robots*, 116–117.

29 Including revolutionary guerrilla movements, if one accepts the argument made by Che Guevara in *Guerrilla Warfare*, trans. J. P. Morray (New York: Vintage Books, 1961).

30 In this respect, Arkin's strategy resembles the program of behavior modification imagined in Anthony Burgess's novel *A Clockwork Orange* and the film of the same title based on it.

31 See P. W. Singer, *Wired for War: The Robotics Revolution and Conflict in the Twenty-First Century* (New York: Penguin, 2009), 119.

32 To our knowledge, the only author so far to have warned against both the abandonment of responsibility and the concentration of power it makes possible is David Graeber in *The Utopia of Rules: On Technology, Stupidity, and the Secret Joys of Bureaucracy* (Brooklyn, N.Y.: Melville House, 2015).

33 The consequences of a robot's actions are themselves not directly taken into account in evaluating its behavior; the only thing that matters is that the relevant rule be respected. The consequences of actions governed by the laws of war nonetheless intervene indirectly. More precisely, they intervene in the past, historically, which is to say when laws of war and rules of engagement are formulated or modified. In the present, by contrast, when an action is taking place, they have no weight whatever in determining the morality of an artificial agent's behavior; rule agreement is the only thing that counts.

34 See P. Strawson, "Freedom and resentment," *Proceedings of the British Academy* 48 (1962): 1–25; and Isaiah Berlin, "Two Concepts of Liberty," in *Four Essays on Liberty* (Oxford: Oxford University Press, 1969), 118–172.

35 On the idea of social institutions as machines—the first archetypal machines, which led to the development of what is called "technology"—see Lewis Mum-

ford, *Technics and Human Development*, vol. 1 of *The Myth of the Machine* (New York: Harcourt, Brace and Jovanovich, 1967). On computers in relation to the government, conceived as a machine, see Jon Agar, *The Government Machine: A Revolutionary History of the Computer* (Cambridge, Mass.: MIT Press, 2003).

36 See Graeber, *Utopia of Rules*, especially 149–205.

37 See, for example, B. Hayes-Roth, "An architecture for adaptive intelligent systems," *Artificial Intelligence* 72, nos. 1–2 (1995): 329–365; D. Gilbert, M. Aparicio, B. Atkinson, S. Brady, J. Ciccarino, B. Grosof, P. O'Connor, et al., *IBM Intelligent Agent Strategy—White Paper* (Yorktown Heights, N.Y.: IBM Corp., 1995); P. Maes, "Artificial life meets entertainment: Lifelike autonomous agents," *Communications of the ACM* 38, no. 11 (1995): 108–114; Stuart J. Russell and Peter Norvig, *Artificial Intelligence: A Modern Approach* (Englewood Cliffs, N.J.: Prentice Hall, 1995; 3rd. ed., 2010); and D. C. Smith, A. Cypher, and J. Spohrer, "KidSim: Programming agents without a programming language," *Communications of the ACM* 37, no. 7 (1994): 55–67.

38 The alternative, contemplated by some theorists today, is to say that the artificial agent itself should be considered a fully moral and legal person and should be held responsible for its actions. In this view, if the agent causes an injury to third parties because it acts contrary to the operator's instructions, it is the agent rather than the operator that is legally responsible.

39 This is another reason why it seems difficult to consider artificial agents as moral persons who are legally responsible for their actions.

40 See Charles Perrow, *Normal Accidents: Living with High-Risk Technologies*, 2nd ed. (Princeton, N.J.: Princeton University Press, 1999); and Charles Perrow, *The Next Catastrophe: Reducing Our Vulnerabilities to Natural, Industrial, and Terrorist Disasters*, 2nd ed. (Princeton, N.J.: Princeton University Press, 2011).

41 Also known as "system accidents"; see Perrow, *Normal Accidents*, 3–5, 63–66.

42 See, for example, the Feelix Growing project directed by Lola Cañamero (http://cordis.europa.eu/project/rcn/99190_en.html) or Cynthia Breazeal's Jibo project (http://myjibo.com/).

43 See Paul Dumouchel, "Y a-t-il des sentiments moraux?," *Dialogue* 43, no. 3 (2004): 471–489.

44 See especially the discussion on the opening page of Chapter 3.

45 See Luisa Damiano, *Filosofia della scienza e medicina riabilitativa in dialogo* (forthcoming).

46 See D. J. Feil-Seifer and M. J. Matarić, "Ethical principles for socially assistive robotics," *IEEE Robotics and Automation Magazine* 18, no. 1 (2011): 24–31; A. J. Moon, P. Danielson, and H. F. M. Van der Loos, "Survey-based discussions on

morally contentious applications of interactive robotics," *International Journal of Social Robotics* 4, no. 1 (2011): 77–91; and Laurel D. Riek and Don Howard, "A Code of Ethics for the Human-Robot Interaction Profession," paper delivered at We Robot 2014, Third Annual Conference on Legal and Policy Issues Relating to Robotics, Coral Gables, Fla., April 4–5, 2014, http://robots.law.miami.edu/2014/wp-content/uploads/2014/03/a-code-of-ethics-for-the-human-robot-interaction-profession-riek-howard.pdf.

47   See Feil-Seifer and Matarič, "Ethical principles for socially assistive robotics."

Works Cited

Adams, Fred, and Ken Aizawa. "Defending the Bounds of Cognition." In Menary, ed., *The Extended Mind*, 67–80.

Adolphs, Ralph. "Could a Robot Have Emotions? Theoretical Perspectives from Social Cognitive Neuroscience." In Fellous and Arbib, eds., *Who Needs Emotions?*, 9–25.

Agar, Jon. *The Government Machine: A Revolutionary History of the Computer.* Cambridge, Mass.: MIT Press, 2003.

Airenti, G. "Aux origines de l'anthropomorphisme: Intersubjectivité et théorie de l'esprit." *Gradhiva* 15 (2012): 35–53.

Alexander, W. H., and O. Sporns. "An embodied model of learning, plasticity, and reward." *Adaptive Behavior* 10, nos. 3–4 (2002): 143–159.

Allen, Colin, and Marc Bekkof. *Species of Mind: The Philosophy and Biology of Cognitive Ethology.* Cambridge, Mass.: MIT Press, 1997.

Alterman, Richard, and David Kirsh, eds. *Proceedings of the 25th Annual Meeting of the Cognitive Science Society.* Boston, Mass.: Cognitive Science Society, 2003.

Anders, Gunther. *L'obsolescence de l'homme: Sur l'âme à l'époque de la deuxième révolution industrielle.* Trans. Christophe David. Paris: Ivrea, 2001.

Anderson, M., and S. L. Anderson. "Machine ethics: Creating an ethical intelligent agent." *AI Magazine* 28, no. 4 (2007): 15–26.

Arbib, M. A., and J. M. Fellous. "Emotions: From brain to robot." *Trends in Cognitive Sciences* 8, no. 12 (2004): 554–561.

Arendt, Hannah. *The Human Condition.* Chicago: University of Chicago Press, 1958.

Arkin, Ronald C. *Behavior-Based Robotics.* Cambridge, Mass.: MIT Press, 1998.

——. *Governing Lethal Behavior in Autonomous Robots.* Boca Raton, Fla.: CRC Press, 2009.

"Artificial Intelligence and Life in 2030: One Hundred Year Study on Artificial Intelligence." Report of the 2015 Study Panel, Stanford University, September 2016. https://ai100.stanford.edu/sites/default/files/ai_100_report_0831fnl.pdf.

Asada, M. "Towards artificial empathy: How can artificial empathy follow the developmental pathway of natural empathy?" *International Journal of Social Robotics* 7, no. 1 (2015): 19–33.

Astington, Janet Wilde. "The Developmental Interdependence of Theory of Mind and Language." In Enfield and Levinson, eds., *Roots of Sociality*, 179–206.

Baron, D., and D. C. Long. "Avowals and first-person privilege." *Philosophy and Phenomenological Research* 62, no. 2 (2001): 311–335.

Bateson, Gregory. *Steps to an Ecology of Mind.* San Francisco: Chandler / New York: Ballantine, 1972.

Baum, Eric B. *What Is Thought?* Cambridge, Mass.: MIT Press, 2004.

Bedau, Mark A., and Emily C. Parke, eds. *The Ethics of Protocells: Moral and Social Implications of Creating Life in the Laboratory.* Cambridge, Mass.: MIT Press, 2009.

Beer, Jenay M., Akanksha Prakash, Tracy L. Mitzner, and Wendy A. Rogers. "Understanding Robot Acceptance." Technical Report HFA-TR-1103. Atlanta: Georgia Institute of Technology, School of Psychology, 2011. https://smartech.gatech.edu /bitstream/handle/1853/39672/HFA-TR-1103-RobotAcceptance.pdf.

Berlin, Isaiah. *Four Essays on Liberty.* Oxford: Oxford University Press, 1969.

Berthouze, L., and T. Ziemke. "Epigenetic robotics: Modeling cognitive development in robotic systems." *Connection Science* 15, no. 3 (2003): 147–150.

Bickle, John. "Multiple Realizability." In Edward N. Zalta, ed., *Stanford Encyclopedia of Philosophy* (Spring 2013 ed.). http://plato.stanford.edu/archives/spr2013/entries /multiple-realizability.

Bien, Z. Zenn, and Dimitar Stefanov, eds. *Advances in Rehabilitation Robotics: Human-Friendly Technologies on Movement Assistance and Restoration for People with Disabilities.* Berlin: Springer, 2004.

Billard, Aude, Annalisa Bonfiglio, Giorgio Cannata, Piero Cosseddu, Torbjorn Dahl, Kerstin Dautenhahn, Fulvio Mastrogiovanni et al. "The RoboSkin Project: Challenges and Results." In Padois et al., eds., *Romansy 19*, 351–358.

Bimbenet, Étienne. *L'animal que je ne suis plus.* Paris: Gallimard, 2011.

Bitbol, M., and P. L. Luisi. "Autopoiesis with or without cognition: Defining life at its edge." *Journal of the Royal Society Interface* 1, no. 1 (2004): 99–107.

Bostrom, Nick. *Superintelligence: Paths, Dangers, Strategies.* Oxford: Oxford University Press, 2014.

Braitenberg, Valentino. *Vehicles: Experiments in Synthetic Psychology.* Cambridge, Mass.: MIT Press, 1984.

Breazeal, C. "Emotions and sociable humanoid robots." *International Journal of Human-Computer Studies* 59, no. 1 (2003): 119–155.

———. "Toward sociable robots." *Robotics and Autonomous Systems* 42, nos. 3–4 (2003): 167–175.

Breazeal, Cynthia. "Artificial Interaction between Humans and Robots." In Kelemen and Sosik, eds., *Advances in Artificial Life,* 582–591.

Budrys, Algis. *The Unexpected Dimension.* New York: Ballantine, 1960.

Burton, A. "Dolphins, dogs, and robot seals for the treatment of neurological diseases." *Lancet Neurology* 12, no. 9 (2013): 851–852.

Cabibihan, J.-J., H. Javed, M. Ang, Jr., and S. M. Aljunied. "Why robots? A survey on the roles and benefits of social robots for the therapy of children with autism." *International Journal of Social Robotics* 5, no. 4 (2013): 593–618.

Caillois, Roger. *Le fleuve alphée.* Paris: Gallimard, 1978.

Calvo, Rafael A., Sidney K. D'Mello, Jonathan Gratch, and Arvid Kappas, eds. *Oxford Handbook of Affective Computing.* Oxford: Oxford University Press, 2014.

Cañamero, L. "Bridging the Gap between HRI and Neuroscience in Emotion Research: Robots as Models." Paper presented at the 2014 Human-Robot Interaction [HRI] Workshop, Bielefeld, Germany. http://www.macs.hw.ac.uk/~kl360/HRI2014W/submission/S16.pdf.

——. "Emotion understanding from the perspective of autonomous robots research." *Neural Networks* 18, no. 4 (2005): 445–455.

Cañamero, Lola, and Ruth Aylett, eds. *Animating Expressive Characters for Social Interaction.* Amsterdam: John Benjamins, 2008.

Cañamero, Lola, Pierre-Yves Oudeyer, and Christian Balkenius, eds. *Proceedings of the Ninth International Conference on Epigenetic Robotics.* Lund University Cognitive Studies, vol. 146. Lund: Department of Philosophy, Lund University, 2009.

Čapek, Karel. *R.U.R. (Rossum's Universal Robots).* Trans. Claudia Novack. New York: Penguin, 2004.

Capitan, W. H., and D. D. Merrill, eds. *Art, Mind, and Religion.* Pittsburgh: University of Pittsburgh Press, 1967.

Carrier, Martin, and Alfred Nordmann, eds. *Science in the Context of Application.* Boston Studies in the Philosophy of Science, vol. 274. New York: Springer, 2011.

Carruthers, Peter, and Peter K. Smith, eds. *Theories of Theories of Mind.* Cambridge: Cambridge University Press, 1996.

Carter, C. Sue, I. Izja Lederhendler, and Brian Kirkpatrick, eds. *The Integrative Neurobiology of Affiliation.* Cambridge, Mass.: MIT Press, 1999.

Casati, Roberto. *The Shadow Club: The Greatest Mystery in the Universe—Shadows—and the Thinkers Who Unlocked Their Secrets.* Trans. Abigail Asher. New York: Knopf, 2003.

Cassam, Quassim, ed. *Self-Knowledge.* Oxford: Oxford University Press, 1994.

Ceruti, M., and L. Damiano. "Embodiment enattivo, cognizione e relazione dialogica." *Encyclopaideia* 37, no. 17 (2013): 19–46.

Chalmers, David J. *The Conscious Mind: In Search of a Fundamental Theory.* New York: Oxford University Press, 1996.

Chapell, Jackie, Susannah Thorpe, Nick Hawes, and Aaron Sloman, eds. *Proceedings of the International Symposium on AI-Inspired Biology 2010*. London: Society for the Study of Artificial Intelligence and the Simulation of Behaviour, 2010.

Chemero, Anthony. *Radical Embodied Cognitive Science*. Cambridge, Mass.: MIT Press, 2010.

Chiel, H. J., and R. D. Beers. "The brain has a body: Adaptive behavior emerges from interactions of nervous system, body, and environment." *Trends in Neurosciences* 20, no. 12 (1997): 553–557.

Chiles, James R. *Inviting Disaster: Lessons from the Edge of Technology*. New York: HarperBusiness, 2001.

Clark, A. "An embodied cognitive science?" *Trends in Cognitive Sciences* 3, no. 9 (1999): 345–351.

Clark, A., and D. J. Chalmers. "The extended mind." *Analysis* 58, no. 1 (1998): 10–23.

Clark, Andy. "*Memento's* Revenge: The Extended Mind, Extended." In Menary, ed., *Extended Mind*, 41–66.

———. *Natural-Born Cyborgs: Minds, Technologies, and the Future of Human Intelligence*. New York: Oxford University Press, 2003.

Coeckelbergh, Mark. *Growing Moral Relations: Critique of Moral Status Ascription*. London: Palgrave Macmillan, 2012.

Cohen, John. *Human Robots in Myth and Science*. London: Allen and Unwin, 1966.

Costa, S., H. Lehmann, K. Dautenhahn, B. Robins, and F. Soares. "Using a humanoid robot to elicit body awareness and appropriate physical interaction in children with autism." *International Journal of Social Robotics* 6, no. 4 (2014): 1–14.

Costa, Sandra, Hagen Lehmann, Ben Robins, Kersten Dautenhahn, and Filomena Soares. "Where Is Your Nose? Developing Body Awareness Skills among Children with Autism Using a Humanoid Robot." In *Proceedings of the Sixth International Conference on Advances in Computer-Human Interactions [ACHI 2013]*, 117–122. Copenhagen: IARIA XPS Press, 2013.

Cowdell, Scott, Chris Fleming, and Joel Hodge, eds. *Mimesis, Movies, and Media*. Vol. 3 of *Violence, Desire, and the Sacred*. New York: Bloomsbury Academic, 2015.

Cruse, Holk. "Robotic Experiments on Insect Walking." In Holland and McFarland, eds., *Artificial Ethology*, 122–139.

Damiano, L., P. Dumouchel, and H. Lehmann. "Towards human-robot affective co-evolution: Overcoming oppositions in constructing emotions and empathy." *International Journal of Social Robotics* 7, no. 1 (2015): 7–18.

Damiano, Luisa. "Co-emergencies in life and science: A double proposal for biological emergentism." *Synthese* 185, no. 2 (2012): 273–294.

———. *Filosofia della scienza e medicina riabilitativa in dialogo*. Forthcoming.

————. *Unità in dialogo: Un nuovo stile per la conoscenza*. Milan: Bruno Mondadori, 2009.

Damiano, Luisa, and Lola Cañamero. "Constructing Emotions: Epistemological Groundings and Applications in Robotics for a Synthetic Approach to Emotions." In Chapell et al., eds., *Proceedings of the International Symposium on AI-Inspired Biology 2010*, 20–28.

Damiano, Luisa, and Paul Dumouchel. "Epigenetic Embodiment." In Cañamero et al., eds., *Proceedings of the Ninth International Conference on Epigenetic Robotics*, 41–48.

Damiano, Luisa, Paul Dumouchel, and Hagen Lehmann. "Should Empathetic Social Robots Have Interiority?" In Ge et al., eds., *Social Robotics*, 268–277.

Damiano, Luisa, Antoine Hiolle, and Lola Cañamero. "Grounding Synthetic Knowledge: An Epistemological Framework and Criteria of Relevance for the Synthetic Exploration of Life, Affect, and Social Cognition." In Lenaerts et al., eds., *Advances in Artificial Life*, 200–207.

Dautenhahn, K. "Methodology and themes of human-robot interaction: A growing research field." *International Journal of Advanced Robotics* 4, no. 1 (2007): 103–108.

————. "Socially intelligent robots: Dimensions of human-robot interaction." *Philosophical Transactions of the Royal Society B: Biological Sciences* 362, no. 1480 (2007): 679–704.

Dautenhahn, K., C. L. Nehaniv, M. L. Walters, B. Robins, H. Kose-Bagci, N. A. Mirza, and M. Blow. "KASPAR—A minimally expressive humanoid robot for human-robot interaction research." *Applied Bionics and Biomechanics* 6, nos. 3–4 (2009): 369–397.

Dautenhahn, Kerstin. "Design Spaces and Niche Spaces of Believable Social Robots." In *Proceedings of the 11th IEEE International Workshop on Robot and Human Interactive Communication*, 192–197. New York: Institute of Electrical and Electronics Engineers, 2002.

————. "Robots as Social Actors: AURORA and the Case of Autism." In *Proceedings of the Third International Cognitive Technology Conference*, 359–374. San Francisco, August 1999.

Dautenhahn, Kerstin, and Chrystopher L. Nehaniv, eds. *Imitation in Animals and Artifacts*. Cambridge, Mass.: MIT Press, 2002.

Davidson, Donald. "Knowing one's own mind." *Proceedings and Addresses of the American Philosophical Association* 60, no. 3 (1987): 441–458.

Dennett, Daniel. *Kinds of Minds: Toward an Understanding of Consciousness*. New York: Basic Books, 1996.

Descartes, René. *A Discourse on Method and Related Writings*. Trans. Desmond M. Clarke. London: Penguin, 1999.

———. *Meditations and Other Metaphysical Writings.* Trans. Desmond M. Clarke. London: Penguin, 1998.

———. *Passions of the Soul.* Trans. Stephen H. Voss. Indianapolis: Hackett, 1989.

Dillon, A., and M. G. Morris. "User acceptance of information technology: Theories and models." *Annual Review of Information Science and Technology* 31 (1996): 3–32.

Dixon, Thomas. *From Passions to Emotions: The Creation of a Secular Psychological Category.* Cambridge: Cambridge University Press, 2003.

Dowe, David. "Is Stephen Hawking Right? Could AI Lead to the End of Humankind?" *IFLScience,* December 12, 2014. http://www.iflscience.com/technology/stephen-hawking-right-could-ai-lead-end-humankind.

Dumouchel, Paul. "Agency, Affiliation, and Emotion." In Trappl, ed., *Cybernetics and Systems,* 727–732.

———. *The Barren Sacrifice: An Essay on Political Violence.* Trans. Mary Baker. East Lansing: Michigan State University Press, 2015.

———. *Émotions: Essai sur le corps et le social.* Paris: Les Empêcheurs de Penser en Rond, 1999.

———. *Emozioni: Saggio sul corpo e sul sociale.* Trans. Luisa Damiano. Milan: Medusa, 2008.

———. "Mirrors of Nature: Artificial Agents in Real Life and Virtual Worlds." In Cowdell et al., eds., *Mimesis, Movies, and Media,* 51–60.

———. "Y a-t-il des sentiments moraux?" *Dialogue* 43, no. 3 (2004): 471–489.

Dumouchel, Paul, and Luisa Damiano. "Artificial empathy, imitation, and mimesis." *Ars Vivendi* 1, no. 1 (2001): 18–31.

Dumouchel, Paul, and Jean-Pierre Dupuy, eds. *L'auto-organisation: De la physique au politique.* Paris: Seuil, 1983.

Ekman, Paul, and Wallace V. Friesen. *Unmasking the Face: A Guide to Recognizing Emotions from Facial Clues.* Englewood Cliffs, N.J.: Prentice-Hall, 1975.

Enfield, N. J., and Stephen C. Levinson, eds. *Roots of Sociality: Culture, Cognition and Interaction.* Oxford: Berg, 2006.

Falconer, Jason. "Panasonic's Robotic Bed / Wheelchair First to Earn Global Safety Certification." *New Atlas,* April 15, 2014. http://www.gizmag.com/panasonic-reysone-robot-bed-wheelchair-is013482/31656/.

Farrow, Tom F. D., and Peter W. R. Woodruff, eds. *Empathy in Mental Illness.* Cambridge: Cambridge University Press, 2007.

Feil-Seifer, D. J., and M. J. Matarić. "Ethical principles for socially assistive robotics." *IEEE Robotics and Automation Magazine* 18, no. 1 (2011): 24–31.

Feil-Seifer, David, and Maja J. Matarić. "Defining Socially Assistive Robotics." In *Proceedings of the 2005 IEEE 9th International Conference on Rehabilitation Robotics [ICORR]*, 465–468. Piscataway, N.J.: Institute of Electrical and Electronics Engineers, 2005.

Fellous, Jean-Marc, and Michael A. Arbib, eds. *Who Needs Emotions? The Brain Meets the Robot.* New York: Oxford University Press, 2005.

Floridi, L., and J. W. Sanders. "On the morality of artificial agents." *Minds and Machines* 14, no. 3 (2004): 349–379.

Fodor, J. A. "Special sciences (or: The disunity of science as a working hypothesis)." *Synthese* 28, no. 2 (1974): 97–115.

Fodor, Jerry A. *The Language of Thought.* New York: Thomas Crowell, 1975.

———. *The Modularity of Mind: An Essay on Faculty Psychology.* Cambridge, Mass.: MIT Press, 1983.

Fong, T., I. Nourbakhsh, and K. Dautenhahn. "A survey of socially interactive robots." *Robotics and Autonomous Systems* 42, nos. 3–4 (2003): 146–166.

Froese, T., and T. Ziemke. "Enactive artificial intelligence: Investigating the systemic organization of life and mind." *Artificial Intelligence* 173, nos. 3–4 (2009): 466–500.

Gallagher, Shaun. "Interpretations of Embodied Cognition." In Tschacher and Bergomi, eds., *The Implications of Embodiment*, 59–70.

Gallese, V. "Intentional attunement: A neurophysiological perspective on social cognition and its disruption in autism." *Brain Research: Cognitive Brain Research* 1079, no. 1 (2006): 15–24.

Gallese, Vittorio. "'Being Like Me': Self-Other Identity, Mirror Neurons, and Empathy." In Hurley and Chater, eds., *Perspectives on Imitation*, 1:101–118.

———. "The 'Shared Manifold' Hypothesis: Embodied Simulation and Its Role in Empathy and Social Cognition." In Farrow and Woodruff, eds., *Empathy in Mental Illness*, 448–472.

Ge, Shuzhi Sam, Oussama Khatib, John-John Cabibihan, Reid Simmons, and Mary-Anne Williams, eds. *Social Robotics: Fourth International Conference [ICSR 2012].* Lecture Notes in Artificial Intelligence, vol. 7621. Berlin: Springer, 2012.

Gibbard, Allan. *Wise Choices, Apt Feelings: A Theory of Normative Judgment.* Cambridge, Mass.: Harvard University Press, 1990.

Gibbs, Raymond W., Jr. *Embodiment and Cognitive Science.* New York: Cambridge University Press, 2000.

———. "Intentions as Emergent Products of Social Interactions." In Malle et al., eds., *Intentions and Intentionality*, 105–122.

Gibbs, Samuel. "Elon Musk: Artificial Intelligence Is Our Biggest Existential Threat." *Guardian*, October 27, 2014.

Gilbert, D., M. Aparicio, B. Atkinson, S. Brady, J. Ciccarino, B. Grosof, P. O'Connor, et al. *IBM Intelligent Agent Strategy—White Paper.* Yorktown Heights, N.Y.: IBM Corp., 1995.

Girard, René. *Violence and the Sacred.* Trans. Patrick Gregory. Baltimore: Johns Hopkins University Press, 1977.

Gontier, Thierry. "Descartes et les animaux-machines: Une réhabilitation?" In Guichet, ed., *De l'animal-machine à l'âme des machines,* 25–44.

Graeber, David. *The Utopia of Rules: On Technology, Stupidity, and the Secret Joys of Bureaucracy.* Brooklyn, N.Y.: Melville House, 2015.

Grasso, Frank. "How Robotic Lobsters Locate Odour Sources in Turbulent Water." In Holland and McFarland, eds., *Artificial Ethology,* 47–59.

Grimaud, Emmanuel, and Zaven Paré. *Le jour où les robots mangeront des pommes: Conversations avec un Geminoïd.* Paris: Petra, 2011.

Guevara, Ernesto. *Guerrilla Warfare.* Trans. J. P. Morray. New York: Vintage Books, 1961.

Guichet, Jean-Luc, ed. *De l'animal-machine à l'âme des machines: Querelles biomécaniques de l'âme, XVIIᵉ–XXIᵉ siècles.* Paris: Publications de la Sorbonne, 2010.

Gunkel, David J. *The Machine Question: Critical Perspectives on AI, Robots, and Ethics.* Cambridge, Mass.: MIT Press, 2012.

Hayes-Roth, B. "An architecture for adaptive intelligent systems." *Artificial Intelligence* 72, nos. 1–2 (1995): 329–365.

Hayles, N. Catherine. *How We Became Posthuman: Virtual Bodies in Cybernetics, Literature, and Informatics.* Chicago: University of Chicago Press, 1999.

Heerink, M., B. Kröse, V. Evers, and B. J. Wielinga. "The influence of social presence on acceptance of a companion robot by older people." *Journal of Physical Agents* 2, no. 2 (2008): 33–40.

Hegel, Frank, Torsten Spexard, Britta Wrede, Gernot Horstmann, and Thurid Vogt. "Playing a Different Imitation Game: Interaction with an Empathetic Android Robot." In *Sixth IEEE-RAS International Conference on Humanoid Robots,* 56–61. Piscataway, N.J.: Institute of Electrical and Electronics Engineers, 2006.

Hillman, Michael. "Rehabilitation Robotics from Past to Present—A Historical Perspective." In Bien and Stefanov, eds., *Advances in Rehabilitation Robotics,* 25–44.

Hobbes, Thomas. *Human Nature; or, The Fundamental Elements of Policy.* Part 1 of *The Elements of Law, Natural and Politic.* Ed. J. C. A. Gaskin. Oxford: Oxford University Press, 1994.

———. *Leviathan; or, The Matter, Forme, and Power of a Commonwealth Ecclesiasticall and Civil.* Ed. Michael Oakeshott. London: Collier Macmillan, 1962.

Holland, Owen, and David McFarland, eds. *Artificial Ethology*. Oxford: Oxford University Press, 2001.

Höök, K. "Affective loop experiences: Designing for interactional embodiment." *Philosophical Transactions of the Royal Society B: Biological Sciences* 364, no. 12 (2009): 3585–3595.

Höök, Kristina. "Affective Loop Experiences—What Are They?" In Oinas-Kukkonen et al., eds., *Persuasive Technology*, 1–12.

Humphreys, Paul. *Extending Ourselves: Computational Science, Empiricism, and Scientific Method*. New York: Oxford University Press, 2004.

Hurley, Susan, and Nick Chater, eds. *Perspectives on Imitation: From Neuroscience to Social Science*. 2 vols. Cambridge, Mass.: MIT Press, 2005.

Iacono, Iolanda, Hagen Lehmann, Patrizia Marti, Ben Robins, and Kerstin Dautenhahn. "Robots as Social Mediators for Children with Autism—A Preliminary Analysis Comparing Two Different Robotic Platforms." In *IEEE International Conference on Development and Learning [ICDL 2011]*, 2 vols., 2:1–6. Piscataway, N.J.: Institute of Electrical and Electronics Engineers, 2011.

Iriki, A., M. Tanaka, and Y. Iwamura. "Coding of modified body schema during tool use by macaque postcentral neurones." *Neuroreport* 7 (1996): 2325–2330.

Izard, Carroll E. "Cross-Cultural Perspectives on Emotion and Emotion Communication." In Triandis and Lonner, eds., *Basic Processes*, 185–220.

Johnson, Deborah G., and Jameson M. Wetmore, eds. *Technology and Society: Building Our Sociotechnical Future*. Cambridge, Mass.: MIT Press, 2009.

Joy, Bill. "Why the Future Doesn't Need Us." *Wired*, April 2000. http://www.wired.com/2000/04/joy-2/.

Karwowski, Waldemar, ed. *International Encyclopedia of Ergonomics and Human Factors*. 3 vols. London: Taylor and Francis, 2001.

Kelemen, Jozef, and Petr Sosik, eds. *Advances in Artificial Life (ECAL 2001)*. Lecture Notes in Computer Science, vol. 2159. Heidelberg: Springer, 2001.

Kim, Jaegwon. *Supervenience and Mind: Selected Philosophical Essays*. New York: Cambridge University Press, 1993.

Krishnan, Armin. *Killer Robots: Legality and Ethicality of Autonomous Weapons*. Burlington, Vt.: Ashgate, 2009.

Krueger, J. "Extended cognition and the space of social interaction." *Consciousness and Cognition* 20, no. 3 (2011): 643–657.

Latour, Bruno. *Petites leçons de sociologie des sciences*. Paris: La Découverte, 2006.

———. "Where Are the Missing Masses? The Sociology of a Few Mundane Artifacts." In Johnson and Wetmore, eds., *Technology and Society*, 151–180.

Le Bihan, Denis. *Looking Inside the Brain*. Trans. T. Lavender Fagan. Princeton, N.J.: Princeton University Press, 2015.

Lehmann, H., I. Iacono, K. Dautenhahn, P. Marti, and B. Robins. "Robot companions for children with Down Syndrome: A case study." *Interaction Studies* 15, no. 1 (2014): 99–112.

Lehmann, Hagen, Iolanda Iacono, Ben Robins, Patrizia Marti, and Kerstin Dautenhahn. " 'Make It Move': Playing Cause and Effect Games with a Robot Companion for Children with Cognitive Disabilities." In *Proceedings of the 29th Annual European Conference on Cognitive Ergonomics [ECCE 2011]*, 105–122. New York: Association for Computing Machinery, 2011.

Lenaerts, Tom, Mario Giacobini, Hugues Bersini, Paul Bourgine, Marco Dorigo, and René Doursat, eds. *Advances in Artificial Life, ECAL 2011*. Cambridge, Mass.: MIT Press, 2011.

Lenhard, Johannes, and Eric Winsberg. "Holism and Entrenchment in Climate Model Validation." In Carrier and Nordmann, eds., *Science in the Context of Application*, 115–130.

Levesque, Hector, and Gerhard Lakemeyer. "Cognitive Robotics." In van Harmelen et al., eds., *Handbook of Knowledge Representation*, 869–886.

Lewis, Marc D., and Isabela Granic, eds. *Emotion, Development, and Self-Organization: Dynamic Systems Approaches to Emotional Development*. New York: Cambridge University Press, 2002.

Li, H., J.-J. Cabibihan, and Y. K. Tan. "Towards an effective design of social robots." *International Journal of Social Robotics* 3, no. 4 (2011): 333–335.

Lin, Patrick, Keith Abney, and George A. Bekey, eds. *Robot Ethics: The Ethical and Social Implications of Robotics*. Cambridge, Mass.: MIT Press, 2012.

Luhmann, Niklas. *A Sociological Theory of Law*. Trans. Elizabeth King and Martin Albrow. London: Routledge and Kegan Paul, 1985.

Lungarella, Max, Fumiya Iida, Josh Bongard, and Rolf Pfeifer, eds. *50 Years of Artificial Intelligence*. Lecture Notes in Computer Science, vol. 4850. Berlin: Springer, 2007.

Lurz, Robert W. *The Philosophy of Animal Minds*. Cambridge: Cambridge University Press, 2009.

MacDorman, K. F., and H. Ishiguro. "The uncanny advantage of using androids in cognitive and social science research." *Interactive Studies* 7, no. 3 (2006): 297–337.

Maes, P. "Artificial life meets entertainment: Lifelike autonomous agents." *Communications of the ACM* 38, no. 11 (1995): 108–114.

Malafouris, Lambros. *How Things Shape the Mind: A Theory of Material Engagement*. Cambridge, Mass.: MIT Press, 2013.

Malamoud, C. "Machines, magie, miracles." *Gradhiva* 15, no. 1 (2012): 144–161.

Malle, Bertram F., Louis J. Moses, and Dare A. Baldwin, eds. *Intentions and Intentionality: Foundations of Social Cognition*. Cambridge, Mass.: MIT Press, 2001.

Manghi, Sergio. "Legame emozionale, legame sociale." Preface to Dumouchel, *Emozioni,* 5–14.

Marr, David. *Vision: A Computational Investigation into the Human Representation and Processing of Visual Information.* San Francisco: W. H. Freeman, 1982.

Maturana, Humberto R., and Francisco J. Varela. *Autopoiesis and Cognition: The Realization of the Living.* Boston Studies in Philosophy of Science, vol. 42. Dordrecht: D. Reidel, 1980.

———. *De máquinas y seres vivos: Autopoiesis, la organización de lo vivo.* Santiago: Editorial Universitaria, 1973; 6th ed., 2004.

———. *The Tree of Knowledge: The Biological Roots of Human Understanding.* Trans. Robert Paolucci. Boston: Shambhala / New Science Library, 1987; rev. ed., 1992.

McFarland, David. *Guilty Robots, Happy Dogs: The Question of Alien Minds.* Oxford: Oxford University Press, 2008.

McGonigle, Brendan. "Robotic Experiments on Complexity and Cognition." In Holland and McFarland, eds., *Artificial Ethology,* 210–224.

Menary, Richard, ed. *The Extended Mind.* Cambridge, Mass.: MIT Press, 2010.

Moon, A. J., P. Danielson, and H. F. M. Van der Loos. "Survey-based discussions on morally contentious applications of interactive robotics." *International Journal of Social Robotics* 4, no. 1 (2011): 77–91.

Moor, J. H. "The nature, importance, and difficulty of machine ethics." *IEEE Intelligent Systems* 21, no. 4 (2006): 18–21.

Moore, R. K. "A Bayesian explanation of the 'uncanny valley' effect and related psychological phenomena." *Nature Science Reports* 2, rep. no. 864 (2012).

Morgan, Mary S., and Margaret Morrison, eds. *Models as Mediators: Perspectives on Natural and Social Science.* Cambridge: Cambridge University Press, 1999.

Mori, M. "Bukimi no tani" [The Uncanny Valley]. *Energy* 7, no. 4 (1970): 33–35.

Morse, Anthony F., Robert Lowe, and Tom Ziemke. "Towards an Enactive Cognitive Architecture." In *Proceedings of the International Conference on Cognitive Systems [CogSys 2008],* Karlsruhe, Germany, April 2008. www.ziemke.org/morse-lowe-ziemke-cogsys2008/.

Mumford, Lewis. *Technics and Human Development.* Vol. 1 of *The Myth of the Machine.* New York: Harcourt, Brace and Jovanovich, 1967.

Nehaniv, Chrystopher L., and Kerstin Dautenhahn, eds. *Imitation and Social Learning in Robots, Humans, and Animals: Behavioural, Social, and Communicative Dimensions.* Cambridge: Cambridge University Press, 2007.

Noë, Alva. *Out of Our Heads: Why You Are Not Your Brain, and Other Lessons from the Biology of Consciousness.* New York: Hill and Wang, 2009.

Nørskov, Marco. *Social Robots: Boundaries, Potential, Challenges.* London: Routledge, 2016.

Núñez, Rafael, and Walter J. Freeman, eds. *Reclaiming Cognition: The Primacy of Action, Intention, and Emotion.* Thorverton, Devon, U.K.: Imprint Academic, 2000.

O'Connor, T. P. *Animals as Neighbors: The Past and Present of Commensal Species.* East Lansing: Michigan State University Press, 2013.

Oinas-Kukkonen, Harri, Per Hasle, Marja Harjumaa, Katarina Segerståhl, and Peter Øhrstrøm, eds. *Persuasive Technology: Proceedings of Third International Conference.* Lecture Notes in Computer Science, vol. 5033. Berlin: Springer, 2008.

O'Mathúna, Dónal P. *Nanoethics: Big Ethical Issues with Small Technology.* London: Continuum, 2009.

Padois, Vincent, Philippe Bidaud, and Oussama Khatib, eds. *Romansy 19—Robot Design, Dynamics, and Control.* CISM International Centre for Mechanical Sciences Series, vol. 544. Vienna: Springer, 2013.

Paiva, Ana, ed. *Affective Interactions: Towards a New Generation of Computer Interfaces.* Lecture Notes in Computer Science / Lecture Notes in Artificial Intelligence, vol. 1814. Berlin: Springer, 2000.

Paiva, Ana, Iolanda Leite, and Tiago Ribeiro. "Emotion Modeling for Social Robots." In Calvo et al., eds., *Oxford Handbook of Affective Computing,* 296–308.

Parisi, D. "Internal robotics." *Connection Science* 16, no. 4 (2004): 325–338.

Parisi, Domenico. *Future Robots: Towards a Robotic Science of Human Beings.* London: John Benjamins, 2014.

Pask, Gordon. *Conversation, Cognition, and Learning: A Cybernetic Theory and Methodology.* Amsterdam: Elsevier, 1975.

Perrow, Charles. *The Next Catastrophe: Reducing Our Vulnerabilities to Natural, Industrial, and Terrorist Disasters.* 2nd ed. Princeton, N.J.: Princeton University Press, 2011.

———. *Normal Accidents: Living with High-Risk Technologies.* 2nd ed. Princeton, N.J.: Princeton University Press, 1999.

Pfeifer, Rolf, and Alexandre Pitti. *La révolution de l'intelligence du corps.* Paris: Manuella, 2012.

Pfeifer, Rolf, and Christian Scheier. *Understanding Intelligence.* Cambridge, Mass.: MIT Press, 1999.

Phelps, E. A. "Emotion and cognition: Insights from studies of the human amygdala." *Annual Review of Psychology* 57 (2006): 27–53.

Piaget, Jean. *Biology and Knowledge: An Essay on the Relations between Organic Regulations and Cognitive Processes.* Trans. Beatrix Walsh. Chicago: University of Chicago Press, 1971.

————. *The Child's Conception of the World.* Trans. Joan Tomlinson and Andrew Tomlinson. London: Routledge and Kegan Paul, 1929.

Pinel, Philippe. *A Treatise on Insanity.* Trans. D. D. Davis. Sheffield, U.K.: W. Todd, 1806.

Proust, Joëlle. *Comment l'esprit vient aux bêtes: Essai sur la représentation.* Paris: Gallimard, 1997.

Putnam, Hilary. "Psychological Predicates." In Capitan and Merrill, eds., *Art, Mind, and Religion,* 37–48.

Pyers, Jennie E. "Constructing the Social Mind: Language and False-Belief Understanding." In Enfield and Levinson, eds., *Roots of Sociality,* 207–228.

Qiao Liang and Wang Xiangsui. *Unrestricted Warfare: China's Master Plan to Destroy America.* Panama City: Pan American, 2002.

Rawlinson, Kevin. "Microsoft's Bill Gates Insists AI Is a Threat." *BBC News Technology,* January 29, 2015. http://www.bbc.com/news/31047780.

Rick, L. D. "Wizard of Oz studies in HRI: A systematic review and new reporting guidelines." *Journal of Human-Robot Interaction* 1, no. 1 (2012): 119–136.

Riek, Laurel D., and Don Howard. "A Code of Ethics for the Human-Robot Interaction Profession." Paper delivered at We Robot 2014, Third Annual Conference on Legal and Policy Issues Relating to Robotics, Coral Gables, Fla., April 4–5, 2014. http://robots.law.miami.edu/2014/wp-content/uploads/2014/03/a-code-of -ethics-for-the-human-robot-interaction-profession-riek-howard.pdf.

Rizzolatti, G., L. Fogassi, and V. Gallese. "Neurological mechanisms underlying the understanding and imitation of action." *Nature Reviews Neuroscience* 2, no. 9 (2001): 661–670.

Rizzolatti, Giacomo, and Corrado Sinigaglia. *Mirrors in the Brain: How Our Minds Share Actions and Emotions.* Trans. Frances Anderson. Oxford: Oxford University Press, 2008.

Rose, Nikolas. *The Politics of Life Itself: Biomedicine, Power, and Subjectivity in the Twenty-First Century.* Princeton, N.J.: Princeton University Press, 2007.

Ross, Don, and Paul Dumouchel. "Emotions as strategic signals." *Rationality and Society* 16, no. 3 (2004): 251–286.

Rowlands, Mark. *The New Science of the Mind: From Extended Mind to Embodied Phenomenology.* Cambridge, Mass.: MIT Press, 2010.

Russell, Stuart J., and Peter Norvig. *Artificial Intelligence: A Modern Approach.* Englewood Cliffs, N.J.: Prentice Hall, 1995; 3rd ed., 2010.

Sandini, Giulio, Giorgio Metta, and David Vernon. "The *iCub* Cognitive Humanoid Robot: An Open-System Research Platform for Enactive Cognition." In Lungarella et al., eds., *50 Years of Artificial Intelligence,* 358–369.

Sartre, Jean-Paul. *L'être et le néant: Essai d'ontologie phénoménologique.* Paris: Gallimard, 1943.

Schneewind, J. B. *The Invention of Autonomy: A History of Modern Moral Philosophy.* Cambridge: Cambridge University Press, 1998.

Sen, Amartya. *The Idea of Justice.* Cambridge, Mass.: Belknap Press of Harvard University Press, 2009.

Shapiro, L. A. "Multiple realizations." *Journal of Philosophy* 97, no. 12 (2000): 635–654.

Shapiro, Lawrence A. *The Mind Incarnate.* Cambridge, Mass.: MIT Press, 2004.

Sharkey, A., and N. Sharkey. "The crying shame of robot nannies: An ethical appraisal." *Interaction Studies* 11, no. 2 (2010): 161–190.

———. "Granny and the robots: Ethical issues in robot care for the elderly." *Ethics and Information Technology* 14, no. 1 (2012): 27–40.

Shibata, Takanori, Kazuyoshi Wada, Tomoko Saito, and Kazuo Tanie. "Psychological and Social Effects to Elderly People by Robot-Assisted Activity." In Cañamero and Aylett, eds., *Animating Expressive Characters for Social Interaction,* 177–193.

Silberstein, M., and A. Chemero. "Complexity and extended phenomenological-cognitive systems." *Topics in Cognitive Science* 4, no. 1 (2012): 35–50.

Simondon, Gilbert. *Du mode d'existence des objets techniques.* Paris: Aubier, 1958.

Singer, P. W. *Wired for War: The Robotics Revolution and Conflict in the Twenty-First Century.* New York: Penguin, 2009.

Slaby, Jan. "Emotions and Extended Mind." In von Scheve and Salmela, eds., *Collective Emotions,* 32–46.

Smith, D. C., A. Cypher, and J. Spohrer. "KidSim: Programming agents without a programming language." *Communications of the ACM* 37, no. 7 (1994): 55–67.

Sparrow, R., and L. Sparrow. "In the hands of machines? The future of aged care." *Minds and Machines* 16, no. 2 (2006): 141–161.

Spier, Emmet. "Robotic Experiments on Rat Instrumental Learning." In Holland and McFarland, eds., *Artificial Ethology,* 189–209.

Steel, Daniel P. *Across the Boundaries: Extrapolation in Biology and Social Science.* New York: Oxford University Press, 2008.

Stewart, John, Olivier Gapenne, and Ezequiel A. Di Paolo, eds. *Enaction: Toward a New Paradigm for Cognitive Science.* Cambridge, Mass.: MIT Press, 2010.

Strawson, P. "Freedom and resentment." *Proceedings of the British Academy* 48 (1962): 1–25.

Thompson, E., and F. J. Varela. "Radical embodiment: Neural dynamics and consciousness." *Trends in Cognitive Sciences* 5, no. 10 (2001): 418–425.

Thompson, Evan. *Mind in Life: Biology, Phenomenology, and Sciences of Mind.* Cambridge, Mass.: Belknap Press of Harvard University Press, 2007.

Trappl, Robert, ed. *Cybernetics and Systems: Proceedings of 11th European Meeting on Cybernetics and Systems Research [EMCSR].* Vienna: Austrian Society for Cybernetics Studies, 2002.

Triandis, Harry C., and Walter J. Lonner, eds. *Basic Processes.* Vol. 3 of *Handbook of Cross-Cultural Psychology.* Boston: Allyn and Bacon, 1980.

Tschacher, Wolfgang, and Claudia Bergomi, eds. *The Implications of Embodiment: Cognition and Communication.* Exeter, U.K.: Imprint Academic, 2011.

Turkle, Sherry. *Alone Together: Why We Expect More from Technology and Less from Each Other.* New York: Basic Books, 2011.

Van Harmelen, Frank, Vladimir Lifschitz, and Bruce Porter, eds. *Handbook of Knowledge Representation.* Amsterdam: Elsevier, 2008.

Varela, Francisco J. *Principles of Biological Autonomy.* New York: North Holland, 1979.

Varela, Francisco J., Evan Thompson, and Eleanor Rosch. *The Embodied Mind: Cognitive Science and Human Experience.* Cambridge, Mass.: MIT Press, 1991.

Vasmatzidis, Ioannis. "User Requirements in Information Technology." In Karwowski, ed., *International Encyclopedia of Ergonomics and Human Factors,* 1:750–753.

Vernon, David. *Artificial Cognitive Systems. A Primer.* Cambridge, Mass.: MIT Press, 2014.

Von Scheve, Christian, and Mikko Salmela, eds. *Collective Emotions.* Oxford: Oxford University Press, 2014.

Wada, K., and T. Shibata. "Living with seal robots—Its sociopsychological and physiological influences on the elderly at a care house." *IEEE Transactions on Robotics* 23, no. 5 (2007): 972–980.

Wada, K., T. Shibata, M. Asada, and T. Musha. "Robot therapy for prevention of dementia at home—Results of preliminary experiment." *Journal of Robotics and Mechatronics* 19, no. 6 (2007): 691–697.

Wainer, J., K. Dautenhahn, B. Robins, and F. Amirabdollahian. "A pilot study with a novel setup for collaborative play of the humanoid robot KASPAR with children with autism." *International Journal of Social Robotics* 6, no. 1 (2014): 45–65.

Wallach, Wendell, and Colin Allen. *Moral Machines: Teaching Robots Right from Wrong.* New York: Oxford University Press, 2008.

Webb, B. "Can robots make good models of biological behavior?" *Behavioral and Brain Sciences* 24, no. 6 (2001): 1033–1050.

Winter, Langdon. "Do Artifacts Have Politics?" In Johnson and Wetmore, eds., *Technology and Society,* 209–226.

Wittgenstein, Ludwig. *Philosophical Investigations.* Trans. G. E. M. Anscombe. 3rd ed. Oxford: Basil Blackwell, 1968.

Worms, Frédéric. *Penser à quelqu'un.* Paris: Flammarion, 2014.

Ziemke, T., and R. Lowe. "On the role of emotion in embodied cognitive architectures: From organisms to robots." *Cognitive Computation* 1, no. 1 (2009): 104–117.

Ziemke, Tom. "What's That Thing Called Embodiment?" In Alterman and Kirsh, eds., *Proceedings of the 25th Annual Meeting of the Cognitive Science Society*, 1305–1310.

Acknowledgments

The inquiry that led to this book began in 2007, when Luisa Damiano obtained a fellowship from the Japan Society for the Promotion of Science (JSPS) to research, under Paul Dumouchel's supervision, new perspectives on empathy in Cognitive Sciences and AI. In 2008 we began a two-year common research project, also funded by JSPS, on the operationalization of empathy in the emerging domains of Social, Cognitive, and Developmental Robotics. To designate the subject of our new research, we coined the label "Artificial Empathy," which is becoming the accepted term for scientific investigations linked to the design and creation of embodied artificial agents endowed with affective competences that promote social interactions with humans. Since 2009, our Artificial Empathy Project has been funded by different institutions: the Imitatio Foundation (San Francisco, USA), the University of Bergamo (Italy), and the Cariplo Foundation (Milan, Italy). We wish to thank all these institutions, as well as the Japan Society for the Promotion of Science, for their financial support. The project is still ongoing. *Living with Robots* is its first important outcome. The book, which was initially published in French by Le Seuil in 2016, reflects nine years of shared research and intense dialogue. Although we elaborated its plan and contents together, each one is responsible for the final form of different parts of the book: Paul Dumouchel for the Introduction and Chapters 1, 2, and 5; and Luisa Damiano for the Preface and Chapters 3 and 4. In September 2016 the French version of the book was awarded the Grand Prix by the Manpower Group Foundation / HEC Paris.

We would like to thank all those who, over the years, in one way or another, helped us in our work, particularly Shin Abiko, Minoru Asada, Gianluca Bocchi, Lola Cañamero, Angelo Cangelosi, Mauro

Ceruti, Luc Faucher, Giuseppe Gembillo, Marcel Hénaff, Hiroshi Ishiguro, Benoît Jacquet, Bill Johnsen, Frédéric Keck, Hiroashi Kitano, Hagen Lehmann, Giulia Lombardi, Trevor Merrill, Giorgio Metta, Zaven Paré, Giulio Sandini, Takanori Shibata, Stefano Tomelleri, as well as the research centers Ars Vivendi at Ritsumeikan University and CERCO at the University of Bergamo. We are grateful to Jean-Pierre Dupuy, generous with both his time and his criticisms as editor of the series in which this book originally appeared (La couleur des idées, Éditions du Seuil, Paris), and for having encouraged and supported it from the first.

Our gratitude goes to Malcolm DeBevoise for his great work in translating the book into English. Last but not least, we would like to sincerely thank Jeff Dean, Executive Editor for Physical Sciences and Technology at Harvard University Press, whose constant and insightful support has been fundamental for the realization of this project.

Credits

p. 135   Sage. Illah R. Nourbakhsh, The Robotics Institute, Carnegie Mellon University, Pittsburgh, PA.

p. 136   Kismet. Computer Science and Artificial Intelligence Laboratory (CSAIL), Massachusetts Institute of Technology. Current installation at the MIT Museum, Cambridge, MA. Photograph © Sam Ogden, Sam Ogden Photography, Newton, MA.

Index

absence, 37–39
action, 169, 191–195
action at a distance, 150–153
adaptability, 35–37
Adolphs, Ralph, 125
affect. *See* emotions and affect
affective coordination, 20, 138–144,
  146–148, 200–202, 224n9
affective loops, 130–132, 141–142, 201
Afghanistan, 187
agency: of artificial agents, 197–198,
  233n38; legal sense of, 197, 233n38; of
  military robots and autonomous
  weapons, 84–85, 177; of mind, 82–88,
  91, of semiautonomous robots, 213n31
Aibo, 1
Allen, Colin, 189
Ambo, Phie, *Mechanical Love*, 163,
  228n30
analytic (partial) agents, 170–171, 195–199
Anders, Günther, 30
animals: cognitive abilities of, 66–70;
  emotions of, 19; minds of, 16, 58–66,
  95; robotic modeling of mind / behavior of, 58–66. *See also* artificial
  animal companions; human-animal
  relations; pets
animats, 58, 156
anime. *See* manga and anime
anthropocentrism, 78, 85
anthropomorphism, 20, 106, 108–111, 120
Arendt, Hannah, 14, 169
Aristotle, 33, 76, 97–99
Arkin, Ronald, 179–186, 189–193, 232n25
art, emotion in, 112

artificial agents: ambivalence toward,
  199; behavior management of, 187–191;
  characteristics of, 170–171; as
  computational entities, 195–196;
  decision making powers ceded to,
  180–186, 193–195, 198–199; ethics of,
  171, 191–192, 230n6, 233n38; regulation
  of, 194–195; substitutes vs., 170–171;
  ubiquity of, 194–195. *See also* analytic
  (partial) agents; robots
artificial animal companions, 34,
  153–158
artificial empathy, 21, 129, 163, 187, 202
artificial ethology, 16, 58 66, 95
artificial intelligence, 71, 122
artificial social agents. *See* social robots
Ashimo, 1
Asimov, Isaac, 175
assemblages, 35
Astro Boy, 6–7, 9
attunement, 141
authority: power vs., 42–43; spurious
  examples of, 41–43; of substitutes,
  31–34, 43
autistic children, therapeutic robots for,
  113, 115, 116, 118, 136, 158–161, 166–167,
  209n5, 228n33
automated devices and systems, 2,
  41–46
autonomous weapons. *See* military
  robots and autonomous weapons
autonomy: action and, 191–195; of
  analytic (partial) agents, 195–199;
  conceptions of, 47–52, 212n26;
  contextual and relational character of,